MONOGRAPHS ON
STATISTICS AND APPLIED PROBABILITY

General Editors

D.R. Cox, D.V. Hinkley, N. Reid, D.B. Rubin and B.W. Silverman

(Full details concerning this series are available from the Publishers).

Multidimensional Scaling

TREVOR F. COX

Department of Mathematics and Statistics
University of Newcastle upon Tyne
UK

and

MICHAEL A.A. COX

School of Business Management
University of Newcastle upon Tyne
UK

CHAPMAN & HALL

London · Glasgow · Weinheim · New York · Tokyo · Melbourne · Madras

Published by Chapman & Hall, 2-6 Boundary Row, London SE1 8HN, UK

Chapman & Hall, 2-6 Boundary Row, London SE1 8HN, UK

Blackie Academic & Professional, Wester Cleddens Road, Bishopbriggs, Glasgow G64 2NZ, UK

Chapman & Hall GmbH, Pappelallee 3, 69469 Weinheim, Germany

Chapman & Hall USA, One Penn Plaza, 41st Floor, New York, NY10119, USA

Chapman & Hall Japan, ITP-Japan, Kyowa Building, 3F, 2-2-1 Hirakawacho, Chiyoda-ku, Tokyo 102, Japan

Chapman & Hall Australia, Thomas Nelson Australia, 102 Dodds Street, South Melbourne, Victoria 3205, Australia

Chapman & Hall India, R. Seshadri, 32 Second Main Road, CIT East, Madras 600 035, India

First edition 1994

© 1994 Trevor F. Cox and Michael A.A. Cox

Printed in Great Britain by St Edmundsbury Press, Bury St Edmunds, Suffolk.

ISBN   0 412 49120 6

A catalogue record for this book is available from the British Library

Library of Congress Catalog Card Number: 94-72654

∞ Printed on permanent acid-free text paper, manufactured in accordance with ANSI/NISO Z39.48-1992 and ANSI/NISO Z39.48-1984 (Permanence of Paper).

To both our families

Contents

Preface

Multidimensional scaling covers a variety of techniques encompassed in the area of multivariate data analysis. In the main the development of the theory of multidimensional scaling has rested in the hands of mathematical psychologists with the journal *Psychometrika* championing the publication of articles in the subject. Multidimensional scaling is now becoming more popular and is extending into areas other than its traditional place in the behavioural sciences. As statistical computer packages include multidimensional scaling in their repetoire, the application of the methods of multidimensional scaling will become even more widespread.

In writing this book we have tried to cover several areas, giving some but not all of the mathematics underlying the theories. The theories have been applied to interesting data sets, hopefully giving insight into the data and the application of the theories. Included with the book is a suite of computer programs that will allow readers to try the techniques for themselves.

Finally we would like to thank the many authors who have contributed to the theory of multidimensional scaling – not just the giants of the subject, Benzécri, Carroll, Coombs, de Leeuw, Gower, Greenacre, Guttman, Heiser, Kruskal, Ramsay, Schönemann, Shepard, Sibson, Takane, Torgerson and Young, but every one of them. For without them this book would not exist.

Newcastle upon Tyne
February 1994

Trevor F. Cox
Michael A.A. Cox

CHAPTER 1

Introduction

1.1 Introduction

Suppose a set of n objects is under consideration and between each pair of objects (r, s) there is a measurement δ_{rs} of the "dissimilarity" between the two objects. For example the set of objects might be ten bottles of whisky each one from a different distillery . The dissimilarity δ_{rs} might be an integer score between zero and ten given to the comparison of the rth and sth whiskies by an expert judge of malt whisky. The judge would be given a tot from the rth bottle and one from the sth and then score the comparison: 0–the whiskies are so alike he cannot tell the difference, to 10– the whiskies are totally different. The judge is presented with all forty-five possible pairs of whiskies and after a pleasant day's work provides the data analyst with a total set of dissimilarities $\{\delta_{rs}\}$. Indeed Lapointe and Legendre (1994) understand the importance of a proper statistical comparison of whiskies, using data from a connoisseur's guide to malt whiskies written by Jackson (1989).

A narrow definition of multidimensional scaling (often abbreviated to MDS) is the search for a low dimensional space, usually Euclidean, in which points in the space represent the objects (whiskies), one point representing one object, and such that the distances between the points in the space, $\{d_{rs}\}$, match as well as possible the original dissimilarities $\{\delta_{rs}\}$. The techniques used for the search for the space and the associated configuration of points form metric and nonmetric multidimensional scaling.

An example
A classic way to illustrate multidimensional scaling is to use journey times between a set of cities in order to reconstruct a map of the cities. Greenacre and Underhill (1982) use flying times between Southern African airports, Mardia *et al.* (1979) use road distances between some British cities.

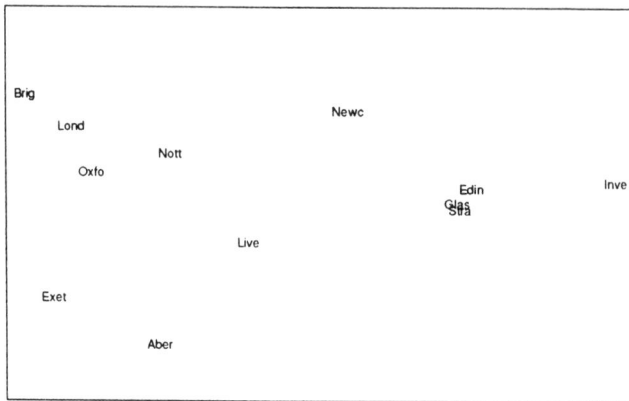

Figure 1.1 *A map of British cities reconstituted from journey time by road. ABER - Aberystwyth, BRIG - Brighton, EDIN - Edinburgh, EXET - Exeter, GLAS - Glasgow, INVE - Inverness, LIVE - Liverpool, LOND - London, NEWC - Newcastle, NOTT - Nottingham, OXFO - Oxford, STRA - Strathclyde.*

For illustration here, the journey times by road between twelve British cities were subjected to multidimensional scaling using classical scaling which is described fully in Chapter 2. Figure 1.1 shows the configuration of points produced by the technique. There is a striking similarity between the positions of the points representing the cities and the positions of the same cities seen in a geographical map of Great Britain, except of course the cities in Figure 1.1 appear to be reflected about a line and rotated from the geographical map usually presented in an atlas.

Multidimensional scaling is not solely about reconstructing maps but can be used on a wide range of dissimilarities arising from various situations, as for example the whisky tasting experiment.

A wider definition of multidimensional scaling can subsume several techniques of multivariate data analysis. At the extreme it covers any technique which produces a graphical representation of objects from multivariate data. For example the dissimilarities obtained from the whisky comparisons could be used in a cluster analysis to find groups of similar whiskies. This text does not attempt to cover all these possibilities as there are many books covering multivariate data analysis in general, for example Mardia *et al.* (1979), Chatfield and Collins (1980), and Krzanowski (1988). The aim here is to give an account of the main topics that could be said to constitute the theory of multidimensional scaling.

Much of the theory of multidimensional scaling was developed in the behavioural sciences, with *Psychometrika* publishing many papers on the subject. It is a tribute to the journal that multidimensional scaling techniques are becoming a popular method of data analysis, with major statistical software packages now incorporating them into their repertoire.

1.2 A look at data and models

Several types of data lend themselves to analysis by multidimensional scaling. Behavioural scientists have adopted several terms relating to data which often are not familiar to others.

1.2.1 Types of data

Variables can be classified according to their "measurement scale". The four scales are the nominal scale, the ordinal scale, the interval scale and the ratio scale.

Nominal scale
Data measured on the nominal scale are classificatory and only different classes are distinguishable, for example hair colour, eye colour.

Ordinal scale
Data on the ordinal scale can be ordered but are not quantitative data. For instance whisky from bottle number 3 might be judged to be of better quality than that from bottle number 7.

Interval scale
Quantitative data where the difference between two values is meaningful are measured on the interval scale. For example temperature in degrees Celsius, the difference in pulse rate before and after exercise.

Ratio scale

Data measured on the ratio scale are similar to those on the interval scale except that the scale has a meaningful zero point, for example weight, height, temperature recorded in Kelvin.

Multidimensional scaling is carried out on data relating objects, individuals, subjects or stimuli to one another. These four terms will often be used interchangeably, although objects obviously refer to inanimate things such as bottles of whisky, individuals and subjects usually refer to people or animals, while stimuli usually refer to non-tangible entities such as the taste of a tot of whisky.

The most common measure of the relationship of one object (stimulus, etc.) to another is a proximity measure. This measures the "closeness" of one object to another, and can either be a "similarity" measure where the similarity of one object to another, s_{rs}, is measured, or a "dissimilarity" measure where the dissimilarity, δ_{rs}, between the two objects is measured.

Suppose for the whisky tasting exercise, several more judges are brought in and each one of them compares all the pairs of whiskies. Then the available data are $\delta_{rs,i}$ where r, s refer to the bottles of whisky, and i refers to the ith judge. The situation now comprises a set of whiskies (stimuli) and a set of judges (subjects).

Number of modes

Each set of objects that underlie the data for multidimensional scaling is called a mode. Thus the dissimilarities $\delta_{rs,i}$ from the whisky tasting above are two-mode data, one-mode being the whiskies and the other the judges.

Number of ways

Each index in the measurement between objects etc. is called a way. So the $\delta_{rs,i}$ above are three-way data.

Thus data for multidimensional scaling are described by their number of modes and their number of ways. With only one whisky judge the data are one-mode, two-way, which is the commonest form. The entries in a two-way contingency table form two-mode, two-way data. An appropriate method of analysis is correspondence analysis described in Chapter 8. Another form of two-mode, two-way data, is where n judges each rank m stimuli. These data can be subjected to unfolding analysis described in Chapter 7. The two-mode, three-way data obtained from the judges of whisky can be analysed by individual differences models of Chapter 9. Three-mode, three-way, or even higher-mode and -way data can be analysed by using some of the methods described in Chapter 11. Data with large number of ways and modes are not very common in practice.

Coombs (1964) gives a classification of types of data. This was updated by Carroll and Arabie (1980) who classify data and also classify types of multidimensional scaling analyses. In so doing they have constructed a useful review of the area. Other useful reviews have been given by Greenacre and Underhill (1982), de Leeuw and Heiser (1982), Wish and Carroll (1982), Gower (1984) and Mead (1992). An introductory book on multidimensional scaling is Kruskal and Wish (1978). Fuller accounts of the subject are given by Schiffman *et al.* (1981), Davidson (1983) and Young (1987) among others.

This book attempts to cover the main constituents of multidimensional scaling giving much but not all of the mathematical theory. Also included in the book is a computer diskette enabling the reader to try out some of the techniques. Instructions for loading the diskette and running the programs are given in the final chapter.

1.2.2 *Multidimensional scaling models*

Some models used for multidimensional scaling are outlined before fuller definition and development in later chapters. The starting point is one-mode, two-way proximity data, and in particular dissimilarity measurements.

Suppose a set of n objects have dissimilarities $\{\delta_{rs}\}$ measured between all pairs of objects. A configuration of n points representing the objects is sought in a p dimensional space. Each point represents one object, with the rth point representing object r. Let the distances, not necessarily Euclidean, between pairs of points be $\{d_{rs}\}$. Then as stated before, the aim of multidimensional scaling is to find a configuration such that the distances $\{d_{rs}\}$ "match" as well as possible the dissimilarities $\{\delta_{rs}\}$. It is the different notions of "matching" that give rise to the different techniques of multidimensional scaling.

Classical scaling
If the distances in the configuration space are to be Euclidean and

$$d_{rs} = \delta_{rs} \qquad r, s = 1, \ldots, n \qquad (1.1)$$

so that the dissimilarities are precisely Euclidean distances, then it is possible to find a configuration of points ensuring the equality in (1.1). Classical scaling treats dissimilarities $\{\delta_{rs}\}$ directly as Euclidean distances and then makes use of the spectral decomposition of a doubly centred matrix of dissimilarities. The technique is discussed fully in Chapter 2.

Least squares scaling
Least squares scaling matches distances $\{d_{rs}\}$ to transformed dissimilarities $\{f(\delta_{rs})\}$, where f is a continuous monotonic function. The function f attempts to transform the dissimilarities into distances whereupon a configuration is found by fitting its associated distances by least squares to $\{f(\delta_{rs})\}$. For example a configuration may be sought such that the loss function

$$\frac{\sum_r \sum_s (d_{rs} - (\alpha + \beta \delta_{rs}))^2}{\sum_r \sum_s d_{rs}^2}$$

is minimized.

Classical scaling and least squares scaling are examples of "metric scaling", where metric refers to the type of transformation of the dissimilarities and not the space in which a configuration of points is sought. Critchley's intermediate method (Critchley, 1978) is another example of metric scaling and is also described in the second chapter.

Nonmetric scaling
If the metric nature of the transformation of the dissimilarities is abandoned, nonmetric multidimensional scaling is arrived at. The transformation f can now be arbitrary but must obey the monotonicity constraint

$$\delta_{rs} < \delta_{r's'} \Rightarrow f(\delta_{rs}) \leq f(\delta_{r's'}) \qquad \text{for all } 1 \leq r, s, r', s' \leq n.$$

Thus only the rank order of the dissimilarities has to be preserved by the transformation and hence the term nonmetric. Nonmetric multidimensional scaling is discussed in Chapter 3.

Procrustes analysis

Suppose multidimensional scaling has been carried out on some dissimilarity data using two different methods giving rise to two configurations of points representing the same set of objects. A Procrustes analysis dilates, translates, reflects and rotates one of the configuration of points to match as well as possible the other, enabling a comparison of the two configurations to be made. Procrustes analysis is covered in Chapter 5.

Unfolding

Suppose n judges of m types of whisky each rank the whiskies in order of their personal preference. Unfolding attempts to produce a configuration of points in a space with each point representing one of the judges together with another configuration of points in the same space, these points representing the whiskies. The configurations are sought so that the rank order of the distances from the ith "judge" point to the "whisky" points, matches as well as possible the original whisky rankings of the ith judge. This is to be the case for all of the judges. Unfolding analysis is the subject of Chapter 7.

Individual differences

Again if m judges each compare all pairs of whiskies, then either m separate multidimensional scaling analyses can be carried out or an attempt can be made at a combined approach. Individual differences models produce an overall configuration of points representing the whiskies in what is called the group stimulus space, together with a configuration of points representing the judges in a different space called the subject space. The position of a particular judge in the subject space depends on the weights needed on the axes of the stimulus space to transform the configuration of points in the group stimulus space into the configuration that would have been peculiar to that judge. Individual differences models are the subject of Chapter 9.

Correspondence analysis

Data in the form of a two-way contingency table can be analysed by correspondence analysis. A space is found in which row profiles can be displayed, and another space for the column profiles. Full discussion is given in Chapter 8.

1.3 Proximities

Proximity literally means nearness in space, time or in some other way. The "nearness" of objects, individuals, stimuli needs definition and measurement prior to statistical analysis. In some situations this is straightforward, but in others difficult and controversial. Measures of proximity are of two types: similarity and dissimilarity with the obvious interpretation of measuring how similar or dissimilar objects are to each other.

Let the objects under consideration comprise a set O. The similarity/dissimilarity measure between two objects is then a real function defined on $O \times O$, giving rise to similarity s_{rs}, or dissimilarity δ_{rs} between the rth and sth objects. Usually $\delta_{rs} \geq 0$, $s_{rs} \geq 0$, and the dissimilarity of an object with itself is taken to be zero, i.e. $\delta_{rr} = 0$. Similarities are usually scaled so that the maximum similarity is unity, with $s_{rr} = 1$.

Hartigan (1967) gives twelve possible proximity structures, S, that might need to be considered before a particular proximity measure is chosen. These are listed in Cormack (1971) and also below.

S1   S defined on $O \times O$ is Euclidean distance

S2   S defined on $O \times O$ is a metric

S3   S defined on $O \times O$ is symmetric real-valued

S4   S defined on $O \times O$ is real-valued

S5   S is a complete ordering $\preceq$ on $O \times O$

S6   S is a partial ordering $\preceq$ on $O \times O$

S7   S is a tree τ on O (a partial similarity order $(r, s) \preceq (r', s')$ whenever $\sup_\tau(r, s) \geq \sup_\tau(r', s')$, see Hartigan or Cormack for further details)

S8   S is a complete relative similarity ordering $\preceq_r$ on O; for each r in O, $s \preceq_r t$ means s is no more similar to r than t is

S9   S is a partial relative similarity order $\preceq_r$ on O

S10 S is a similarity dichotomy on $O \times O$ in which $O \times O$ is divided into a set of similar pairs and a set of dissimilar pairs

S11 S is a similarity trichotomy on $O \times O$ consisting of similar pairs, dissimilar pairs, and the remaining pairs

S12 S is a partition of O into sets of similar objects

Structure S1 is a very strict structure with dissimilarity defined as Euclidean distance. Relaxing this to the requirement of a metric gives S2, where it is recalled that δ_{rs} is a metric if

$$\delta_{rs} = 0 \qquad \text{if and only if } r = s,$$
$$\delta_{rs} = \delta_{sr} \qquad \text{for all } 1 \leq r, s \leq n,$$
$$\delta_{rs} \leq \delta_{rt} + \delta_{ts} \qquad \text{for all } 1 \leq r, s, t \leq n.$$

Relaxing the metric requirement to δ_{rs} being symmetric real-valued or real-valued gives structures S3 and S4. Losing ratio/interval scales of measurement of δ_{rs} leads to the nonmetric structures S5 to S12. Of these the highest structure, S5, has a complete ordering of the $\{\delta_{rs}\}$. The lowest structure, S12, simply partitions O into sets of similar objects.

Choice of proximity measure depends upon the problem at hand, and is often not an easy task. Sometimes similarity between two objects is not based on any underlying data recorded on the objects. For example in the whisky tasting exercise, the judge simply uses taste and smell sensations to produce a score between zero and ten. The similarity/dissimilarity measurement is totally subjective. It is extremely unlikely that the dissimilarities arrived at by the judge would obey proximity structure S1 since they are all integer-valued. The only possibility would be if the whiskies could be represented by integer points on a one dimensional Euclidean space and differences between points generated all forty-five dissimilarities correctly. It is even unlikely that S2 would be satisfied. The most likely structure is S3 or possibly S5 if actual scores were ignored and only the rank order of the dissimilarities taken into account.

In other situations similarities (dissimilarities) are constructed from a data matrix for the objects. These are then called similarity (dissimilarity) coefficients. Several authors, for example Cormack (1971), Jardine and Sibson (1971), Anderberg (1973), Sneath and Sokal (1973), Diday and Simon (1976), Mardia *et al.* (1979), Gordon (1981a,b), Gower (1985), Gower and Legendre (1986), Digby and Kempton (1987), Jackson *et al.* (1989), Snijders *et al.* (1990) discuss various similarity and dissimilarity measures together with their associated problems. The following synthesis of their work

Table 1.1 *Dissimilarity measures for quantitative data*

Euclidean distance	$\delta_{rs} = \left\{ \sum_i (x_{ri} - x_{si})^2 \right\}^{\frac{1}{2}}$
Weighted Euclidean	$\delta_{rs} = \left\{ \sum_i w_i (x_{ri} - x_{si})^2 \right\}^{\frac{1}{2}}$
Mahalanobis distance	$\delta_{rs} = \{ (\mathbf{x}_r - \mathbf{x}_s)^T \boldsymbol{\Sigma}^{-1} (\mathbf{x}_r - \mathbf{x}_s) \}^{\frac{1}{2}}$
City block metric	$\delta_{rs} = \sum_i |x_{ri} - x_{si}|$
Minkowski metric	$\delta_{rs} = \left\{ \sum_i |x_{ri} - x_{si}|^\lambda \right\}^{\frac{1}{\lambda}} \quad \lambda \geq 1$
Canberra metric	$\delta_{rs} = \sum_i |x_{ri} - x_{si}| / (x_{ri} + x_{si})$
Bray-Curtis	$\delta_{rs} = \dfrac{1}{p} \dfrac{\sum_i |x_{ri} - x_{si}|}{\sum_i (x_{ri} + x_{si})}$
Bhattacharyya distance	$\delta_{rs} = \left\{ \sum_i (x_{ri}^{\frac{1}{2}} - x_{si}^{\frac{1}{2}})^2 \right\}^{\frac{1}{2}}$
Angular separation	$\delta_{rs} = \dfrac{\sum_i x_{ri} x_{si}}{[\sum_i x_{ri}^2 \sum_i x_{si}^2]^{\frac{1}{2}}}$
Correlation	$\delta_{rs} = 1 - \dfrac{\sum_i (x_{ri} - \bar{x}_r)(x_{si} - \bar{x}_s)}{\left\{ \sum_i (x_{ri} - \bar{x}_r)^2 \sum_i (x_{si} - \bar{x}_s)^2 \right\}^{\frac{1}{2}}}$

attempts to outline the main ideas behind forming dissimilarities from a data matrix.

Let $\mathbf{X} = [x_{ri}]$ denote the data matrix obtained for n objects on p variables. The vector of observations for the rth object is denoted by $\mathbf{x}_r$, and so $\mathbf{X} = [\mathbf{x}_r^T]$.

Quantitative data
Table 1.1 gives a list of possible dissimilarity measures for quantitative data that are in particular, continuous, possibly discrete, but not binary.

Table 1.2 *Similarity coefficients for binary data*

Czekanowski, Sørensen, Dice	$s_{rs} = \dfrac{2a}{2a + b + c}$
Hamman	$s_{rs} = \dfrac{a - (b + c) + d}{a + b + c + d}$
Jaccard coefficient	$s_{rs} = \dfrac{a}{a + b + c}$
Kulezynski	$s_{rs} = \dfrac{a}{a + b}$
Mountford	$s_{rs} = \dfrac{2a}{a(b + c) + 2bc}$
Mozley, Margalef	$s_{rs} = \dfrac{a(a + b + c + d)}{(a + b)(a + c)}$
Ochiai	$s_{rs} = \dfrac{a}{[(a + b)(a + c)]^{\frac{1}{2}}}$
Phi	$s_{rs} = \dfrac{ad - bc}{[(a + b)(a + c)(b + d)(c + d)]^{\frac{1}{2}}}$
Rogers, Tanimoto	$s_{rs} = \dfrac{a + d}{a + 2b + 2c + d}$
Russell, Rao	$s_{rs} = \dfrac{a}{a + b + c + d}$
Simple matching coefficient	$s_{rs} = \dfrac{a + d}{a + b + c + d}$
Yule	$s_{rs} = \dfrac{ad - bc}{ad + bc}$

Binary data
When all the variables are binary it is usual to construct a similarity coefficient and then to transform this into a dissimilarity coefficient. The measure of similarity between objects r and s is based on the following table.

		Object s		
		1	0	
Object r	1	a	b	$a + b$
	0	c	d	$c + d$
		$a+c$	$b+d$	$p = a + b$ $+c + d$

The table shows the number of variables, a, out of the total p variables where both objects score "1", the number of variables, b, where r scores "1" and s scores "0", etc. Table 1.2 gives a list of similarity coefficients based on the four counts a, b, c, d. Various situations call for particular choices of coefficients. In practice more than one can be tried hoping for some robustness against choice.

Nominal and ordinal data

If, for the ith nominal variable, objects r and s share the same categorization let $s_{rsi} = 1$ and 0 otherwise. A similarity measure is then $p^{-1} \sum_i s_{rsi}$. Of course if other information is available regarding the relationship of various categories for the variables, then s_{rsi} can be given an appropriate value. For example if the variable "bottle shape" has categories: standard (st); short cylindrical (sh); tall cylindrical (ta); and square section (sq), the following "agreement scores" may be appropriate for bottles r and s.

		bottle r			
		st	sh	ta	sq
	st	1.0	0.5	0.5	0.0
bottle s	sh	0.5	1.0	0.3	0.0
	ta	0.5	0.3	1.0	0.0
	sq	0.0	0.0	0.0	1.0

So if bottle r is "tall cylindrical" and bottle s "standard" then $s_{rsi} = 0.5$ for example.

If a variable is ordinal with k categories, then $k - 1$ indicator variables can be used to represent these categories. The indicator variables can then be subjected to similarity coefficients in order to give a value to s_{rsi}. For instance if a bottle variable is "height of the bottle" with categories: small; standard; tall; long and thin, then the variable might be categorized as follows.

Indicator variable

category	I_1	I_2	I_3
small	0	0	0
standard	1	0	0
tall	1	1	0
long and thin	1	1	1

If bottle r is "standard" and bottle s is "long and thin", then using the simple matching coefficient to measure similarity for this variable, $s_{rsi}=0.33$. For further details see Sneath and Sokal (1973) or Gordon (1981a).

Mixed data
Gower (1971) introduced a general similarity coefficient

$$s_{rs} = \frac{\sum_{i=1}^{p} \omega_{rsi} s_{rsi}}{\sum_i \omega_{rsi}},$$

where s_{rsi} is the similarity between the rth and sth objects based on the ith variable alone, and ω_{rsi} is unity if the rth and sth objects can be compared on the ith variable and zero if they cannot. Thus s_{rsi} is an average over all possible similarities s_{rsi} for the rth and sth objects. So for example if some data are missing the overall coefficient is comprised of just those observations which are present for both the rth and sth objects.

Gower suggests the following values for s_{rsi} and ω_{rsi} for binary variables measuring presence/absence.

object r	object s	s_{rsi}	ω_{rsi}
+	+	1	1
+	-	0	1
-	+	0	1
-	-	0	0

For nominal variables Gower suggests $s_{rsi} = 1$ if objects r and s share the same categorization for variable i, $s_{rsi} = 0$ otherwise. Of course other measures such as those described in the previous section can be used.

For quantitative variables,

$$s_{rsi} = 1 - |x_{ri} - x_{si}|/R_i,$$

where R_i is the range of the observations for variable i.

Gower's coefficient can be generalized using weights $\{w_i\}$ for the variables to

$$s_{rsi} = \frac{\sum_i s_{rsi}\omega_{rsi}w_i}{\sum_i \omega_{rsi}w_i}.$$

Further discussion, for example Gordon (1981a), can be found on missing values, incompatibility of units of measurement, conditionally present variables and the weighting of variables.

Transforming from similarities to dissimilarities
Often similarity coefficients have to be transformed into dissimilarity coefficients. Possible transformations are

$$\delta_{rs} = 1 - s_{rs}$$

$$\delta_{rs} = c - s_{rs} \text{ for some constant } c$$

$$\delta_{rs} = \{2(1 - s_{rs})\}^{\frac{1}{2}}.$$

Choice will depend on the problem at hand.

1.3.1 The metric nature of dissimilarities

Gower and Legendre (1986) discuss in detail metric and Euclidean properties of many dissimilarity coefficients. A summary is given of some of the important results they establish or report on.

Let the dissimilarities $\{\delta_{rs}\}$ be placed in a matrix $\mathbf{D}$, the dissimilarity matrix. Similarly let similarities $\{s_{rs}\}$ be placed in a similarity matrix $\mathbf{S}$. Then $\mathbf{D}$ is called metric if δ_{rs} is a metric. Matrix $\mathbf{D}$ is also Euclidean if n points can be embedded in a Euclidean space such that the Euclidean distance between the rth and sth points is δ_{rs}, for all $1 \leq r, s \leq n$.

If $\mathbf{D}$ is nonmetric then the matrix with elements $\delta_{rs} + c$ $(r \neq s)$ is metric where $c \geq \max_{i,j,k} |\delta_{ij} + \delta_{ik} - \delta_{jk}|$.

If $\mathbf{D}$ is metric then so are matrices with elements (i) $\delta_{rs} + c^2$ (ii) $\delta_{rs}^{1/\lambda}$ $(\lambda \geq 1)$ (iii) $\delta_{rs}/(\delta_{rs} + c^2)$ for any real constant c, and $r \neq s$.

Let matrix $\boldsymbol{\Lambda} = [-\frac{1}{2}d_{rs}^2]$.

Then $\mathbf{D}$ is Euclidean if and only if the matrix $(\mathbf{I}-\mathbf{1s}^T)\boldsymbol{\Lambda}(\mathbf{I}-\mathbf{s1}^T)$ is positive semi-definite, where $\mathbf{I}$ is the identity matrix, $\mathbf{1}$ is a vector of ones, and $\mathbf{s}$ is a vector such that $\mathbf{s}^T\mathbf{1} = 1$.

If $\mathbf{S}$ is a positive semi-definite similarity matrix with elements $0 \leq s_{rs} \leq 1$ and $s_{rr} = 1$, then the dissimilarity matrix with elements $d_{rs} = (1 - s_{rs})^{\frac{1}{2}}$ is Euclidean.

If $\mathbf{D}$ is a dissimilarity matrix, then there exists a constant h such that the matrix with elements $(\delta_{rs}^2 + h)^{\frac{1}{2}}$ is Euclidean, where $h \geq -\lambda_n$, the smallest eigenvalue of $\boldsymbol{\Lambda}_1 = \mathbf{H}\boldsymbol{\Lambda}\mathbf{H}$, $\mathbf{H}$ being the centring matrix $(\mathbf{I} - \mathbf{11}^T/n)$.

If $\mathbf{D}$ is a dissimilarity matrix, then there exists a constant k such that the matrix with elements $(\delta_{rs} + k)$ is Euclidean, where $k \geq \mu_n$, the largest eigenvalue of

$$\begin{bmatrix} \mathbf{0} & 2\boldsymbol{\Lambda}_1 \\ -\mathbf{I} & -4\boldsymbol{\Lambda}_2 \end{bmatrix}$$

where $\boldsymbol{\Lambda}_2 = [-\frac{1}{2}d_{rs}]$.

These last two theorems give solutions to the additive constant problem which is discussed further in Chapter 2.

For binary variables Gower and Legendre define

$$S_\theta = \frac{a + d}{a + d + \theta(b + c)} \qquad T_\theta = \frac{a}{a + \theta(b + c)}.$$

Then for the appropriate choice of θ similarity coefficients in Table 1.2 can be obtained. Gower and Legendre show:

For $\theta \geq 1$, $1 - S_\theta$ is metric; $\sqrt{1 - S_\theta}$ is metric for $\theta \geq \frac{1}{3}$; if $\theta < 1$ then $1 - S_\theta$ may be nonmetric; if $\theta < \frac{1}{3}$ then $\sqrt{1 - S_\theta}$ may be nonmetric. There are similar results when S_θ is replaced by T_θ.

If $\sqrt{1 - S_\theta}$ is Euclidean then so is $\sqrt{1 - S_\phi}$ for $\phi \geq \theta$, with a similar result for T_θ.

For $\theta \geq 1$, $\sqrt{1 - S_\theta}$ is Euclidean; for $\theta \geq \frac{1}{2}$, $\sqrt{1 - T_\theta}$ is Euclidean. However $1 - S_\theta$ and $1 - T_\theta$ may be non-Euclidean.

Gower and Legendre give a table of various similarity/dissimilarity coefficients and use these results to establish which coefficients are metrics and which are also Euclidean.

1.3.2 Distribution of proximity coefficients

In calculating the similarity between a pair of objects it might be of interest to establish whether the value obtained is significantly different from that expected for two arbitrary objects. This is usually not an easy task since the underlying distribution of the data vector needs to be known. If multivariate normality can be assumed then some progress is possible with the dissimilarity measures in Table 1.1. For example for the Mahalanobis distance, $\delta_{rs} \sim 2\chi_p^2$.

For similarities based on binary variables, Goodall (1967) found the mean and variance for the simple matching coefficient, assuming independence of the variables. For the simple matching coefficient

$$s_{rs} = p^{-1}(a + d) = p^{-1}(I_1 + I_2 + \ldots + I_p),$$

where $I_i = 1$ if objects r and s agree (i.e. are both 0 or 1) on the ith variable. Let $Pr\{X_{ri} = 1\} = p_i$. Then

$$E(s_{rs}) = p^{-1} \sum_{i=1}^{p} (p_i^2 + (1 - p_i)^2) = \mu,$$

and after some algebra

$$\mathrm{Var}(s_{rs}) = p^{-1} \left\{ \mu(1 - \mu) - p^{-1} \sum_{i=1}^{p} \left(p_i^2 + (1 - p_i)^2 - \mu^2 \right)^2 \right\}.$$

These results can be generalized to the case where the objects can come from different groups which have differing p_i's.

Moments for other coefficients which do not have a constant denominator are much more difficult to obtain. Snijders *et al.* (1990) give a brief review of the derivation of the moments for the Jaccard coefficient and the Dice coefficient. They also extend the results to the case where the binary variables are dependent. Approximate distributions can be found using these moments and hence the significance of an observed value of the similarity coefficient can be assessed.

1.3.3 Dissimilarity of variables

Sometimes it is not the objects that are to be subjected to multidimensional scaling but the variables. One possibility for defining dissimilarities for variables is simply to reverse the roles of objects and variables and to proceed regardless, using one of the dissimilarity measures. Another possibility is to choose a dissimilarity more appropriate to variables than objects. The sample correlation coefficient r_{ij} is often used as the basis for dissimilarity among variables. For instance $\delta_{ij} = 1 - r_{ij}$ could be used. This measure has its critics. A similar dissimilarity can be based on the angular separation of the vectors of observations associated with the ith and jth variables,

$$\frac{\sum_r x_{ri} x_{rj}}{(\sum_r x_{ri}^2 \sum_r x_{rj}^2)^{\frac{1}{2}}}$$

Zegers and ten Berge (1985), Zegers (1986) and Fagot and Mazo (1989) consider general similarity coefficients for variables measured on different metric scales. The argument is that the similarity coefficient has to be invariant under admissible transformations of the variables. The scales considered are: the absolute scale where only the identity transformation is possible; the difference scale which is only invariant under additive transformations; the ratio scale which is only invariant under positive multiplicative transformations; and the interval scale which is invariant up to positive linear transformations. The variables are transformed to "uniformity" according to type:

$$u_{ri} = x_{ri} \quad \text{for the absolute scale}$$

$$u_{ri} = x_{ri} - \bar{x}_i \quad \text{for the difference scale}$$

$$u_{ri} = \left(\frac{1}{n}\sum_s x_{si}^2\right)^{-\frac{1}{2}} x_{ri} \quad \text{for the ratio scale}$$

$$u_{ri} = \left(\frac{1}{n-1}\sum_s (x_{si} - \bar{x}_i)^2\right)^{-\frac{1}{2}} (x_{ri} - \bar{x}_i) \quad \text{for the interval scale.}$$

Consider a general similarity coefficient, s_{ij}, based on the mean squared difference,

$$s_{ij} = 1 - cn^{-1}\sum_r (u_{ri} - u_{rj})^2,$$

where c is a constant. This is to have maximum value unity when

$u_{ri} = u_{rj}$ $(1 \leq r \leq n)$. Hence c can be determined from the requirement that $s_{ij} = s_{ji}$, and then after some algebra

$$s_{ij} = \frac{2 \sum_r u_{ri} u_{rj}}{\left(\sum_r u_{ri}^2 + \sum_r u_{rj}^2 \right)}.$$

This can be a considered alternative to the haphazard use of the sample correlation coefficient.

1.4 Matrix results

A review of matrix algebra is not given here as it is assumed that the reader is familiar with such. However a brief reminder is given of the spectral decomposition of a symmetric matrix, the singular value decomposition of a rectangular matrix, and the Moore-Penrose inverse of a matrix. For outlines of matrix algebra relevant to statistics, see Mardia *et al.* (1979), Healy (1986) for example.

1.4.1 The spectral decomposition

Let $\mathbf{A}$ be an $n \times n$ symmetric matrix, with eigenvalues $\{\lambda_i\}$ and associated eigenvectors $\{\mathbf{v}_i\}$, such that $\mathbf{v}_i^T \mathbf{v}_i = 1$ $(i = 1, \ldots, n)$. Then $\mathbf{A}$ can be written

$$\mathbf{A} = \mathbf{V} \mathbf{\Lambda} \mathbf{V}^T = \sum_{i=1}^{n} \lambda_i \mathbf{v}_i \mathbf{v}_i^T,$$

where

$$\mathbf{\Lambda} = \text{diag}(\lambda_1, \ldots, \lambda_n), \quad \mathbf{V} = [\mathbf{v}_1, \ldots, \mathbf{v}_n].$$

Matrix $\mathbf{V}$ is orthonormal, so that $\mathbf{V}\mathbf{V}^T = \mathbf{V}^T\mathbf{V} = \mathbf{I}$. Also if $\mathbf{A}$ is nonsingular

$$\mathbf{A}^m = \mathbf{V} \mathbf{\Lambda}^m \mathbf{V}^T$$

with $\mathbf{\Lambda}^m = \text{diag}(\lambda_1^m, \ldots, \lambda_n^m)$ for any integer m. If the eigenvalues $\{\lambda_i\}$ are all positive then rational powers of $\mathbf{A}$ can be defined in a similar way and in particular for powers $\frac{1}{2}$ and $-\frac{1}{2}$.

1.4.2 The singular value decomposition

If $\mathbf{A}$ is an $n \times p$ matrix of rank r, then $\mathbf{A}$ can be written as

$$\mathbf{A} = \mathbf{U} \mathbf{\Lambda} \mathbf{V}^T,$$

where $\Lambda = \text{diag}(\lambda_1, \lambda_2, \ldots, \lambda_r)$, with $\lambda_1 \geq \lambda_2 \geq \ldots \geq \lambda_r \geq 0$, $\mathbf{U}$ is an orthonormal matrix of order $n \times r$, and $\mathbf{V}$ an orthonormal matrix of order $r \times r$, i.e. $\mathbf{U}^T\mathbf{U} = \mathbf{V}^T\mathbf{V} = \mathbf{I}$. The set of values $\{\lambda_i\}$ are called the singular values of $\mathbf{A}$. If $\mathbf{U}$ and $\mathbf{V}$ are written in terms of their column vectors, $\mathbf{U} = [\mathbf{u}_1, \ldots, \mathbf{u}_r]$, $\mathbf{V} = [\mathbf{v}_1, \ldots, \mathbf{v}_r]$, then $\{\mathbf{u}_i\}$ are the left singular vectors of $\mathbf{A}$ and $\{\mathbf{v}_i\}$ are the right singular vectors. The matrix $\mathbf{A}$ can then be written as

$$\mathbf{A} = \sum_{i=1}^{r} \lambda_i \mathbf{u}_i \mathbf{v}_i^T.$$

It can be shown that $\{\lambda_i^2\}$ are the nonzero eigenvalues of the symmetric matrix $\mathbf{A}\mathbf{A}^T$ and also of the matrix $\mathbf{A}^T\mathbf{A}$. The vectors $\{\mathbf{u}_i\}$ are the corresponding normalized eigenvectors of $\mathbf{A}\mathbf{A}^T$, and the vectors $\{\mathbf{v}_i\}$ are the corresponding normalized eigenvectors of $\mathbf{A}^T\mathbf{A}$.

An example
As an example let

$$\mathbf{A} = \begin{bmatrix} 5 & 2 & 9 \\ 0 & 1 & 2 \\ 2 & 1 & 4 \\ -4 & 3 & 2 \end{bmatrix}$$

Then the SVD of $\mathbf{A}$ is

$$\mathbf{A} = \begin{bmatrix} 0.901 & 0.098 \\ 0.169 & -0.195 \\ 0.394 & 0.000 \\ 0.056 & -0.980 \end{bmatrix} \begin{bmatrix} 11.619 & 0 \\ 0 & 5.477 \end{bmatrix}$$

$$\times \begin{bmatrix} 0.436 & 0.218 & 0.873 \\ 0.802 & -0.535 & -0.267 \end{bmatrix}$$

or equivalently

$$\mathbf{A} = 11.619 \begin{bmatrix} 0.393 & 0.196 & 0.787 \\ 0.074 & 0.037 & 0.148 \\ 0.172 & 0.086 & 0.344 \\ 0.024 & 0.012 & 0.049 \end{bmatrix}$$

$$+ 5.477 \begin{bmatrix} 0.079 & -0.052 & -0.026 \\ -0.156 & 0.104 & 0.052 \\ 0.000 & 0.000 & 0.000 \\ -0.786 & 0.524 & 0.262 \end{bmatrix}$$

If there are no multiplicities within the singular values then the SVD is unique. If k of the singular values are equal, then the SVD is unique only up to arbitrary rotations in the subspaces spanned by the corresponding left and right singular vectors.

Greenacre (1984) gives a good review of the SVD of a matrix and its use in statistical applications. Its usefulness is that it can be used to approximate matrix $\mathbf{A}$ of rank r by matrix

$$\tilde{\mathbf{A}}_{r^\star} = \sum_{i=1}^{r^\star} \lambda_i \mathbf{u}_i \mathbf{v}_i^T$$

which is of rank $r^\star < r$. The approximation is in fact the least squares approximation of $\mathbf{A}$ found by minimizing

$$\sum_i \sum_j (a_{ij} - x_{ij})^2 = \text{tr}\{(\mathbf{A} - \mathbf{X})(\mathbf{A} - \mathbf{X}^T)\},$$

for all matrices $\mathbf{X}$ of rank $r^\star$ or less. For example, with $r^\star = 1$, $\mathbf{A}$ above is approximated by

$$\tilde{\mathbf{A}}_1 = \begin{bmatrix} 4.56 & 2.28 & 9.14 \\ 0.86 & 0.42 & 1.71 \\ 2.00 & 1.00 & 4.00 \\ 0.28 & 0.14 & 0.57 \end{bmatrix}$$

and noting that the second and third columns of $\tilde{\mathbf{A}}_1$ are simply multiples of the first column. This is of course expected since $\tilde{\mathbf{A}}_1$ is of rank one. If $\mathbf{A}$ is viewed as a matrix representing four points in a three dimensional space, it is noted that only two dimensions are in fact needed to represent the points since $\mathbf{A}$ has rank 2. A one dimensional space approximating to the original configuration is given by $\tilde{\mathbf{A}}_1$ giving an ordering of the points as 4,2,3,1.

Note that the singular value decomposition can be defined so that $\mathbf{U}$ is an $n \times n$ matrix, Λ is an $n \times p$ matrix and $\mathbf{V}$ is a $p \times p$ matrix. These matrices are the same as those just defined but contain extra rows/columns of zeros.

Generalized SVD
Suppose now weighted Euclidean distances are used in the spaces spanning the columns and rows of $\mathbf{A}$. Then the generalized SVD of matrix $\mathbf{A}$ is given by

$$\mathbf{A} = \mathbf{U}\Lambda\mathbf{V}^T,$$

where $\boldsymbol{\Lambda} = \mathrm{diag}(\lambda_1, \lambda_2, \ldots, \lambda_r)$, with $\lambda_1 \geq \lambda_2 \geq \ldots \geq \lambda_r \geq 0$, are the generalized singular values of $\mathbf{A}$, $\mathbf{U}$ is an $n \times r$ matrix, orthonormal with respect to $\boldsymbol{\Omega}$, and $\mathbf{V}$ is a $p \times r$ matrix orthonormal with respect to $\boldsymbol{\Phi}$, i.e. $\mathbf{U}^T \boldsymbol{\Omega} \mathbf{U} = \mathbf{V}^T \boldsymbol{\Phi} \mathbf{V} = \mathbf{I}$.

Let $\mathbf{U} = [\mathbf{r}_1, \ldots, \mathbf{r}_n]$ and $\mathbf{V} = [\mathbf{c}_1, \ldots, \mathbf{c}_p]$. The approximation of $\mathbf{A}$ by a lower rank matrix $\tilde{\mathbf{A}}_{r^\star}$ is given by

$$\tilde{\mathbf{A}}_{r^\star} = \sum_{i=1}^{r^\star} \lambda_i \mathbf{u}_i \mathbf{v}_i^T$$

where $\tilde{\mathbf{A}}_{r^\star}$ is the matrix that minimizes

$$\mathrm{tr}\{\boldsymbol{\Omega}(\mathbf{A} - \mathbf{X})\boldsymbol{\Phi}(\mathbf{A} - \mathbf{X})^T\}$$

over all matrices $\mathbf{X}$ of rank $r^\star$ or less.

1.4.3 The Moore-Penrose inverse

Consider the matrix equation

$$\mathbf{AX} = \mathbf{B}$$

where $\mathbf{A}$ is an $n \times p$ matrix, $\mathbf{X}$ is a $p \times n$ matrix and $\mathbf{B}$ is an $n \times n$ matrix. The matrix $\mathbf{x}$ which minimizes the sum of squares $\mathrm{tr}(\mathbf{AX} - \mathbf{B})^T(\mathbf{AX} - \mathbf{B})$ and itself has the smallest value of $\mathrm{tr}\mathbf{X}^T\mathbf{X}$ among all least squares solutions is given by

$$\mathbf{X} = \mathbf{A}^+\mathbf{B},$$

where $\mathbf{A}^+$ is the unique $p \times n$ Moore-Penrose generalized inverse of $\mathbf{A}$, defined by the equations

$$\mathbf{AA}^+\mathbf{A} = \mathbf{A}$$
$$\mathbf{A}^+\mathbf{AA}^+ = \mathbf{A}^+$$
$$(\mathbf{AA}^+)^\star = \mathbf{Aa}^+$$
$$(\mathbf{A}^+\mathbf{A})^\star = \mathbf{A}^+\mathbf{a},$$

$\mathbf{A}^\star$ being the conjugate transpose of $\mathbf{A}$. As only real matrices are used in this book "$\star$" can be replaced by "T" the usual matrix transpose. For further details see Barnett (1990).

CHAPTER 2

Metric multidimensional scaling

2.1 Introduction

Suppose there are n objects with dissimilarities $\{\delta_{rs}\}$. Metric MDS attempts to find a set of points in a space where each point represents one of the objects and the distances between points $\{d_{rs}\}$ are such that

$$d_{rs} \approx f(\delta_{rs}),$$

where f is a continuous parametric monotonic function. The function f can either be the identity function or a function that attempts to transform the dissimilarities to a distance-like form.

Mathematically, let the objects comprise a set O. Let the dissimilarity, defined on O × O, between objects r and s be δ_{rs} $(r, s \ \epsilon \ \text{O})$. Let ϕ be an arbitrary mapping from O to E, where E is usually a Euclidean space, but not necessarily so, in which a set of points are to represent the objects. Thus let $\phi(r) = x_r$ $(r \ \epsilon \ \text{O}, \ x_r \ \epsilon \ \text{E})$, and let X $= \{x_r : r \ \epsilon \ \text{O}\}$, the image set. Let the distance between the points x_r, x_s in X be given by d_{rs}. The aim is to find a mapping ϕ, for which d_{rs} is approximately equal to $f(\delta_{rs})$ for all $r, s \ \epsilon \ \text{O}$.

The two main metric MDS methods, classical scaling and least squares scaling, will be considered in this chapter, with most emphasis placed on the former.

2.2 Classical scaling

Classical scaling originated in the 1930s when Young and Householder (1938) showed how starting with a matrix of distances between points in a Euclidean space, coordinates for the points can be found such that distances are preserved. Torgerson (1952) brought the subject to popularity using the technique for scaling.

2.2.1 Recovery of coordinates

Chapter 1 saw an application of classical scaling where a map of British cities was constructed from journey times by road between the cities. Suppose the starting point for the procedure had been the actual Euclidean distances between the various cities (making the assumption that Great Britain is a two dimensional Euclidean plane). Can the original positions of the cities be found? They can, but only relative to each other since any solution can be translated, rotated and reflected giving rise to another equally valid solution. The method for finding the original Euclidean coordinates from the derived Euclidean distances was first given by Schoenberg (1935) and Young and Householder (1938). It is as follows.

Let the coordinates of n points in a p dimensional Euclidean space be given by $\mathbf{x}_r$ $(r = 1, \ldots, n)$, where $\mathbf{x}_r = (x_{r1}, \ldots, x_{rp})^T$. Then the Euclidean distance between the rth and sth points is given by

$$d_{rs}^2 = (\mathbf{x}_r - \mathbf{x}_s)^T (\mathbf{x}_r - \mathbf{x}_s). \qquad (2.1)$$

Let the inner product matrix be $\mathbf{B}$, where

$$[\mathbf{B}]_{rs} = b_{rs} = \mathbf{x}_r^T \mathbf{x}_s.$$

From the known squared distances $\{d_{rs}\}$, this inner product matrix $\mathbf{B}$ is found, and then from $\mathbf{B}$ the unknown coordinates.

To find $\mathbf{B}$

Firstly, to overcome the indeterminancy of the solution due to arbitrary translation, the centroid of the configuration of points is placed at the origin. Hence

$$\sum_{r=1}^{n} x_{ri} = 0 \qquad (i = 1, \ldots, p).$$

To find $\mathbf{B}$, from (2.1)

$$d_{rs}^2 = \mathbf{x}_r^T \mathbf{x}_r + \mathbf{x}_s^T \mathbf{x}_s - 2\mathbf{x}_r^T \mathbf{x}_s, \qquad (2.2)$$

and hence

$$\frac{1}{n}\sum_{r=1}^{n} d_{rs}^2 = \frac{1}{n}\sum_{r=1}^{n} \mathbf{x}_r^T\mathbf{x}_r + \mathbf{x}_s^T\mathbf{x}_s,$$

$$\frac{1}{n}\sum_{s=1}^{n} d_{rs}^2 = \mathbf{x}_r^T\mathbf{x}_r + \frac{1}{n}\sum_{s=1}^{n} \mathbf{x}_s^T\mathbf{x}_s,$$

$$\frac{1}{n^2}\sum_{r=1}^{n}\sum_{s=1}^{n} d_{rs}^2 = \frac{2}{n}\sum_{r=1}^{n} \mathbf{x}_r^T\mathbf{x}_r. \tag{2.3}$$

Substituting into (2.2) gives

$$b_{rs} = \mathbf{x}_r^T\mathbf{x}_s,$$

$$= -\frac{1}{2}\left(d_{rs}^2 - \frac{1}{n}\sum_{r=1}^{n} d_{rs}^2 - \frac{1}{n}\sum_{s=1}^{n} d_{rs}^2 + \frac{1}{n^2}\sum_{r=1}^{n}\sum_{s=1}^{n} d_{rs}^2\right),$$

$$= a_{rs} - a_{r.} - a_{.s} + a_{..}, \tag{2.4}$$

where $a_{rs} = -\frac{1}{2}d_{rs}^2$, and

$$a_{r.} = n^{-1}\sum_{s} a_{rs}, \quad a_{.s} = n^{-1}\sum_{r} a_{rs}, \quad a_{..} = n^{-2}\sum_{r}\sum_{s} a_{rs}.$$

Define matrix $\mathbf{A}$ as $[\mathbf{A}]_{rs} = a_{rs}$, and hence the inner product matrix $\mathbf{B}$ is

$$\mathbf{B} = \mathbf{HAH} \tag{2.5}$$

where $\mathbf{H}$ is the centring matrix,

$$\mathbf{H} = \mathbf{I} - n^{-1}\mathbf{11}^T,$$

with $\mathbf{1} = (1, 1, \ldots, 1)^T$, a vector of n ones.

To recover the coordinates from $\mathbf{B}$

The inner product matrix, $\mathbf{B}$, can be expressed as

$$\mathbf{B} = \mathbf{XX}^T,$$

where $\mathbf{X} = [\mathbf{x}_1, \ldots, \mathbf{x}_n]^T$ is the $n \times p$ matrix of coordinates. The rank of $\mathbf{B}$, $r(\mathbf{B})$, is then

$$r(\mathbf{B}) = r(\mathbf{XX}^T) = r(\mathbf{X}) = p.$$

Now $\mathbf{B}$ is symmetric, positive semi-definite and of rank p, and hence has p non-negative eigenvalues and $n - p$ zero eigenvalues.

Matrix **B** is now written in terms of its spectral decomposition,

$$\mathbf{B} = \mathbf{V}\boldsymbol{\Lambda}\mathbf{V}^T,$$

where $\boldsymbol{\Lambda} = \text{diag}(\lambda_1, \lambda_2, \ldots, \lambda_n)$, the diagonal matrix of eigenvalues $\{\lambda_i\}$ of **B**, and $\mathbf{V} = [\mathbf{v}_1, \ldots, \mathbf{v}_n]$, the matrix of corresponding eigenvectors, normalized such that $\mathbf{v}_i^T\mathbf{v}_i = 1$. For convenience the eigenvalues of **B** are labelled such that $\lambda_1 \geq \lambda_2 \geq \ldots \geq \lambda_n \geq 0$. Because of the $n - p$ zero eigenvalues, **B** can now be rewritten as

$$\mathbf{B} = \mathbf{V}_1\boldsymbol{\Lambda}_1\mathbf{V}_1^T,$$

where

$$\boldsymbol{\Lambda}_1 = \text{diag}(\lambda_1, \ldots, \lambda_p), \qquad \mathbf{V}_1 = [\mathbf{v}_1, \ldots, \mathbf{v}_p].$$

Hence it can be seen that the coordinate matrix **X** is given by

$$\mathbf{X} = \mathbf{V}_1\boldsymbol{\Lambda}_1^{\frac{1}{2}},$$

where $\boldsymbol{\Lambda}_1^{\frac{1}{2}} = \text{diag}(\lambda_1^{\frac{1}{2}}, \ldots, \lambda_p^{\frac{1}{2}})$, and thus the coordinates of the points have been recovered from the distances between the points. The arbitrary sign of the eigenvectors $\{\mathbf{v}_i\}$ leads to invariance of the solution with respect to reflection in the origin.

2.2.2 Dissimilarities as Euclidean distances

To be of practical use a configuration of points needs to be found for a set of dissimilarities $\{\delta_{rs}\}$ rather than simply for true Euclidean distances between points $\{d_{rs}\}$.

Suppose dissimilarities $\{\delta_{rs}\}$ are used instead of distances d_{rs} to define matrix **A**, which is then doubly centred to produce matrix **B** as just described. Then it is interesting to ask under what circumstances **B** can give rise to a configuration of points in Euclidean space, using the spectral decomposition, so that the associated distances $\{d_{rs}\}$ are such that $d_{rs} = \delta_{rs}$ for all r, s. The answer is that if **B** is positive semi-definite of rank p, then a configuration in p dimensional Euclidean space can be found. Proofs are presented in de Leeuw and Heiser (1982) or Mardia *et al.* (1979).

Following Mardia *et al.*, if **B** is positive semi-definite of rank p, then

$$\mathbf{B} = \mathbf{V}\boldsymbol{\Lambda}\mathbf{V}^T = \mathbf{X}\mathbf{X}^T,$$

where

$$\Lambda = \mathrm{diag}(\lambda_1, \ldots, \lambda_p), \qquad \mathbf{X} = [\mathbf{x}_r]^T, \qquad \mathbf{x}_r = \lambda^{\frac{1}{2}} \mathbf{v}_r.$$

Now the distance between the rth and sth points of the configuration is given by $(\mathbf{x}_r - \mathbf{x}_s)^T (\mathbf{x}_r - \mathbf{x}_s)$, and hence

$$\begin{aligned}
(\mathbf{x}_r - \mathbf{x}_s)^T (\mathbf{x}_r - \mathbf{x}_s) &= \mathbf{x}_r^T \mathbf{x}_r + \mathbf{x}_s^T \mathbf{x}_s - 2\mathbf{x}_r^T \mathbf{x}_s \\
&= b_{rr} + b_{ss} - 2b_{rs} \\
&= a_{rr} + a_{ss} - 2a_{rs} = -2a_{rs} = \delta_{rs}^2,
\end{aligned}$$

by substituting for b_{rs} using equation (2.4). Thus the distance between the rth and sth points in the Euclidean space is equal to the original dissimilarity δ_{rs}.

Incidently the converse is also true that if $\mathbf{B}$ is formed from Euclidean distances then it is positive semi-definite. For if the points originate from a p dimensional Euclidean space, then

$$\mathbf{B} = \mathbf{HAH} = \mathbf{HZZ}^T\mathbf{H},$$

where $\mathbf{A} = \mathbf{ZZ}^T$ since $\mathbf{A}$ is symmetric. Hence $\mathbf{B} = (\mathbf{HZ})(\mathbf{HZ})^T$ which implies that $\mathbf{B}$ is positive semi-definite.

The next question to be asked is how many dimensions are required in general for the configuration of points produced from a positive semi-definite matrix $\mathbf{B}$ of dissimilarities. It is easily shown that $\mathbf{B}$ has at least one zero eigenvalue, since $\mathbf{B1} = \mathbf{HAH1} = \mathbf{0}$. Thus a configuration of points in an $n - 1$ dimensional Euclidean space can always be found whose associated distances are equal to the dissimilarities $\{\delta_{rs}\}$.

If the dissimilarities give rise to a matrix $\mathbf{B}$ which is not positive semi-definite, a constant can be added to all the dissimilarities (except the self-dissimilarities δ_{rr}) which will then make $\mathbf{B}$ positive semi-definite. Thus taking the distances as $d_{rs} = \delta_{rs} + c(1 - \delta_{KR}^{rs})$, where c is an appropriate constant and δ_{KR}^{rs} the Kronecker delta, will make $\mathbf{B}$ positive semi-definite. This is the additive constant problem, see for example Cailliez (1983), which will be explored further in Section 2.2.8. Once $\mathbf{B}$ has been made positive semi-definite a Euclidean space can be found as before where distances d_{rs} are exactly equal to dissimilarities δ_{rs}.

2.2.3 Classical scaling in practice

It was shown in the previous section that a Euclidean space of at most $n-1$ dimensions could be found so that distances in the space equalled original dissimilarities, which were perhaps modified by the addition of a constant. Usually matrix $\mathbf{B}$ used in the procedure will be of rank $n-1$ and so the full $n-1$ dimensions are needed in the space, and hence little has been gained in dimension reduction of the data.

The configuration obtained could be rotated to its principal axes in the principal components sense; i.e. the projections of the points in the configuration onto the first principal axis have maximum variation possible, the projection of the points onto the second principle axis have maximum variation possible, but subject to this second axis being orthogonal to the first axis, etc.. Then only the first p, $(p < n-1)$ axes are chosen for representing the configuration. However this need not be undertaken since the procedure for finding $\mathbf{X}$ already has the points referred to their principal axes. This is easily seen since in searching for the principal axes,

$$\mathbf{X}^T\mathbf{X} = (\mathbf{V}_1\boldsymbol{\Lambda}_1^{\frac{1}{2}})^T(\mathbf{V}_1\boldsymbol{\Lambda}_1^{\frac{1}{2}})$$
$$= \boldsymbol{\Lambda}^{\frac{1}{2}}\mathbf{V}_1^T\mathbf{V}_1\boldsymbol{\Lambda}^{\frac{1}{2}} = \boldsymbol{\Lambda},$$

and $\boldsymbol{\Lambda}$ is a diagonal matrix.

Gower (1966) was the first to state clearly the formulation and the importance of the classical scaling technique, and from this selection of the first p "principal coordinates" for the configuration he coined the name "principal coordinates analysis" (PCO). Principal coordinates analysis is now synonymous with classical scaling, as also is the term metric scaling. However metric scaling encompasses more than this one technique.

Thus in the spectral decomposition of the matrix $\mathbf{B}$, the distances between the points in the $n-1$ dimensional Euclidean space are given by

$$d_{rs}^2 = \sum_{i=1}^{n-1} \lambda_i(x_{ri} - x_{si})^2,$$

and hence if many of the eigenvalues are "small", then their contribution to the squared distance d_{rs}^2 can be neglected. If only p eigenvalues are retained as being significantly large, then the p dimensional Euclidean space formed for the first p eigenvalues and with $\mathbf{x}_r$ truncated to the first p elements can be used to represent the objects. Hopefully p will be small, preferably 2 or 3 for ease of graphical representation.

The selection of these first p principal coordinates is optimal in the following sense when $\{d_{rs}\}$ are Euclidean. If $\mathbf{x}_r^*$ is a projection of $\mathbf{x}_r$ onto a p' dimensional space with $p' \leq p$ and with associated distances between points $\{d_{rs}^*\}$, then it is precisely the projection given by using the first p' principal coordinates that minimizes

$$\sum\sum(d_{rs}^2 - d_{rs}^{*2}).$$

For the non-Euclidean case the above does not hold but Mardia (1978) has given the following optimal property. For the matrix $\mathbf{B} = \mathbf{HAH}$ a positive semi-definite matrix $\mathbf{B}^* = [b_{rs}]$ of rank at most t is sought such that

$$\sum\sum(b_{rs} - b_{rs}^*)^2 = \text{tr}(\mathbf{B} - \mathbf{B}^*)^2$$

is a minimum.

Let $\lambda_1^* \geq \ldots \geq \lambda_n^*$ be the eigenvalues of $\mathbf{B}^*$ and so at least $n - t$ of these must be zero due to the rank constraint. Then

$$\min \text{tr}(\mathbf{B} - \mathbf{B}^*)^2 = \min \sum_{k=1}^n (\lambda_k - \lambda_k^*)^2.$$

For the minimum

$$\lambda_k^* = \max(\lambda_k, 0) \qquad k = 1, \ldots, t$$
$$= 0 \qquad k = t+1, \ldots, n.$$

So if $\mathbf{B}$ has t or more positive eigenvalues then the first t principal coordinates derived from $\mathbf{B}$ are used for the projection. If $\mathbf{B}$ has fewer than t positive eigenvalues then the space of dimension less than t defined by the positive eigenvalues of $\mathbf{B}$ is used.

Hence in practice if it is found that $\mathbf{B}$ is not positive semi-definite (simply by noting whether there are any negative eigenvalues) then there is a choice of procedure. Either the dissimilarites are modified by adding an appropriate constant, or the negative eigenvalues are simply ignored. If the negative eigenvalues are small in magnitude then little is lost. If they are large then some argue that classical scaling is still appropriate as an exploratory data technique for dimension reduction.

2.2.4 How many dimensions?

As indicated above the eigenvalues $\{\lambda_i\}$ indicate how many dimensions are required for representing the dissimilarities $\{\delta_{rs}\}$. If **B** is positive semi-definite then the number of nonzero eigenvalues gives the number of dimensions required. If **B** is not positive semi-definite then the number of positive eigenvalues is the appropriate number of dimensions. These are the maximum dimensions of the space required. However to be of practical value the number of dimensions of the chosen space needs to be small. Since the coordinates recovered by the procedure are referred to their principal coordinates then simply choosing the first p eigenvalues and eigenvectors of **B** $(p = 2$ or 3 say) will give a small dimensional space for the points.

The sum of squared distances between points in the full space is from (2.3)

$$\frac{1}{2}\sum_{r=1}^{n}\sum_{s=1}^{n}d_{rs}^2 = n\sum_{r=1}^{n}\mathbf{x}_r^T\mathbf{x}_r = n\mathrm{tr}\mathbf{B} = n\sum_{r=1}^{n-1}\lambda_i.$$

A measure of the proportion of variation explained by using only p dimensions is

$$\frac{\sum_{i=1}^{p}\lambda_i}{\sum_{i=1}^{n-1}\lambda_i}.$$

If **B** is not positive semi-definite this measure is modified to

$$\frac{\sum_{i=1}^{p}\lambda_i}{\sum_{i=1}^{n-1}|\lambda_i|} \quad \text{or} \quad \frac{\sum_{i=1}^{p}\lambda_i}{\sum(\text{positive eigenvalues})}.$$

Choice of p can then be assessed with this measure.

2.2.5 A practical algorithm for classical scaling

Although the various steps in the algorithm for classical scaling can be gleaned from the text in the previous sections, it is summarized here.

1. Obtain dissimilarities $\{\delta_{rs}\}$.
2. Find matrix $\mathbf{A} = [-\frac{1}{2}\delta_{rs}^2]$.
3. Find matrix $\mathbf{B} = [a_{rs} - a_{r.} - a_{.s} + a_{..}]$.
4. Find the eigenvalues $\lambda_1, \ldots, \lambda_{n-1}$ and associated eigenvectors $\mathbf{v}_1, \ldots, \mathbf{v}_{n-1}$, where the eigenvectors are normalized so that

$\mathbf{v}_i^T \mathbf{v}_i = \lambda_i$. If B is not positive semi-definite (some of the eigenvalues are negative), either (i) ignore the negative values and proceed, or (ii) add an appropriate constant c to the dissimilarities, $\delta'_{rs} = \delta_{rs} + c(1 - \delta_{KR}^{rs})$ (see Section 2.2.7) and return to step 2.

5. Choose an appropriate number of dimensions p. Possibly use $\sum_1^p \lambda_i / \sum$ (positive eigenvalues) for this.

6. The coordinates of the n points in the p dimensional Euclidean space are given by $x_{ri} = v_{ir}$ $(r = 1, \ldots, n; \ i = 1, \ldots, p)$.

2.2.6 A grave example

There is a fascinating paper in the very first volume of *Biometrika*, published in 1901-1902, concerning cranial measurements on an ancient race of people from Egypt. The paper is by Cicely Fawcett (1901) who was assisted by Alice Lee and others, including the legendary K. Pearson and G.U. Yule. The paper is sixty pages long and gives an insight into the problems faced by statisticians of the time who had no access to modern computing facilities or advanced statistical methods.

Till that time little statistical work had been done for, or by, craniologists on skull measurements although several data sets had been collected. Karl Pearson had asked Professor Flinders Petrie to try to obtain one hundred skulls from a homogeneous race when he embarked on his Egyptian expedition in 1894. Professor Petrie managed to get four hundred skulls, together with their skeletons, sent back to University College in London. These were taken from cemeteries of the Naqada race in Upper Egypt, and were dated at about 8000 years old. Karl Pearson was credited as the first person to calculate correlations for length and breadth in skulls, studying modern German, modern French and the Naqada crania. The second study of the Naqada crania was started in 1895 by Karl Pearson's team, and the time taken to carry out extensive hand calculation of means, standard deviations, correlations, skewness, kurtosis, and probability density fitting, delayed publication until 1901-1902. (O for a computer, or even a hand held calculator!)

The Fawcett paper details the method of measuring the skulls, for which various measuring devices had been deployed, such as a craniometer, a goniometer and a Spengler's pointer. In all, forty-eight measurements and indices were taken and and were published at the end of the paper. The statistical analyses used on the data

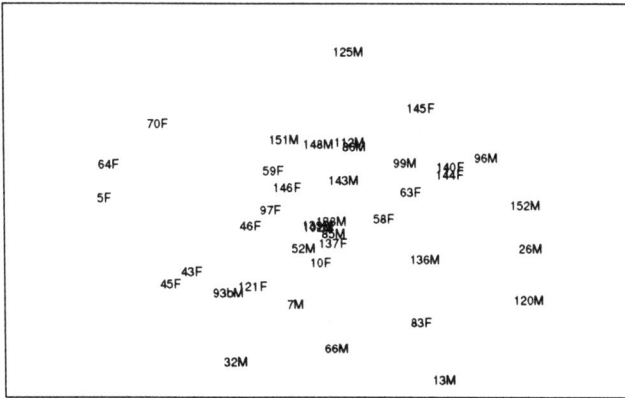

Figure 2.1(i) *Classical scaling of the skull data*

would be classified as basic by modern standards, with means, standard deviations, correlations, etc. being compared in tables. Of course no use could be made of hypothesis tests, confidence intervals, let alone multivariate methods such as cluster analysis, principal components or discriminant analysis.

The results obtained by Karl Pearson's team will not be generally discussed here, since they are rather long and of interest mainly to craniologists. However, to concentrate on one point, the team said it found it impossible to give diagrams of the forty-seven variables, separated by sex. They chose twelve of these variables and constructed twenty-four histograms and fitted density functions, all calculated and plotted at considerable cost of time. The variables were: (i) greatest length, L; (ii) breadth,B; (iii) height, H; (iv) auricular height, OH; (v) circumference above the superciliary ridges, U; (vi) sagittal circumference, S; (vii) cross-circumference, Q; (viii) upper face height, G'H; (ix) nasal breadth, NB; (x) nasal height, NH; (xi) cephalic index, B/L; (xii) ratio of height to length, H/L. These twelve variables will be used for an MDS analysis for 22 male and 18 female skulls. For clarity only a subset of the skulls were chosen and those used had no missing values.

The twelve variables were standardized to have zero mean and

Table 2.1 *The first five leading eigenvalues and eigenvectors of* **B** *giving principal coordinates of the skull data.*

Eigenvalue	Eigenvector
$\lambda_1 = 11.47$	(-1.16, -0.19, -0.07, 0.56, 1.01, -0.49, -0.71, -0.82, -0.42, -0.15, 0.26, -0.30, 0.40, -1.13, 0.02, -0.88, 0.45, 0.00, 0.11, -0.53, 0.79, -0.32, 0.37, -0.08, 0.09, 1.00, -0.41, 0.09, 0.47, 0.00, -0.01, -0.08, 0.60, 0.05, 0.60, 0.45, -0.23, -0.07, -0.24, 0.98)
$\lambda_2 = 4.98$	(0.11, -0.42, 0.21, -0.79, -0.14, -0.70, -0.26, -0.32, -0.03, -0.14, 0.00, 0.24, 0.14, 0.27, -0.64, 0.47, -0.51, -0.07, 0.36, -0.36, 0.31, 0.05, 0.28, -0.04, 0.38, -0.40, -0.33, 0.83, -0.19, -0.12, -0.01, -0.03, 0.26, 0.20, 0.22, 0.55, 0.16, 0.37, 0.40, 0.07)
$\lambda_3 = 4.56$	(-0.12, 0.15, -0.61, -0.10, -0.31, -0.07, -0.21, 0.33, -0.68, -0.01, 0.36, 0.56, -0.26, 0.07, -0.30, -0.16, -0.08, -0.02, -0.18, -0.30, -0.50, -0.69, -0.07, 0.06, 0.65, 0.34, 0.36, -0.25, 0.64, 0.49, 0.18, 0.30, -0.09, -0.02, 0.26, -0.20, 0.27, 0.45, -0.05, -0.19)
$\lambda_4 = 2.55$	(0.16, 0.04, -0.12, -0.12, 0.24, 0.15, 0.04, 0.20, 0.25, -0.16, -0.33, 0.39, 0.48, -0.20, -0.36, -0.07, 0.22, 0.53, -0.18, 0.02, 0.29, -0.55, 0.35, -0.15, -0.32, -0.19, 0.14, 0.10, 0.09, -0.27, 0.24, -0.05, 0.12, -0.09, 0.02, -0.15, -0.24, 0.17, -0.29, -0.44)
$\lambda_5 = 1.73$	(-0.03, -0.09, 0.23, 0.13, 0.07, -0.29, -0.11, 0.43, -0.08, -0.16, -0.04, -0.32, -0.18, 0.19, -0.37, -0.26, 0.32, 0.12, 0.17, 0.24, -0.20, -0.14, 0.11, 0.42, 0.15, -0.20, 0.05, 0.16, 0.06, 0.04, -0.25, -0.22, 0.40, 0.16, -0.25, -0.10, 0.09, -0.13, -0.10, 0.01)

standard deviation unity. Then dissimilarities between skulls were calculated using Euclidean distance and subjected to classical scaling.

Table 2.1 gives the five leading eigenvalues of **B** together with their associated eigenvectors. These are the first five principal coordinates. A two dimensional configuration is obtained by using the first two of these and is shown in Figure 2.1(i). Male and female skulls are marked M and F respectively, and for clarity the points for males and females are given separately in Figures 2.1(ii) and

(ii)

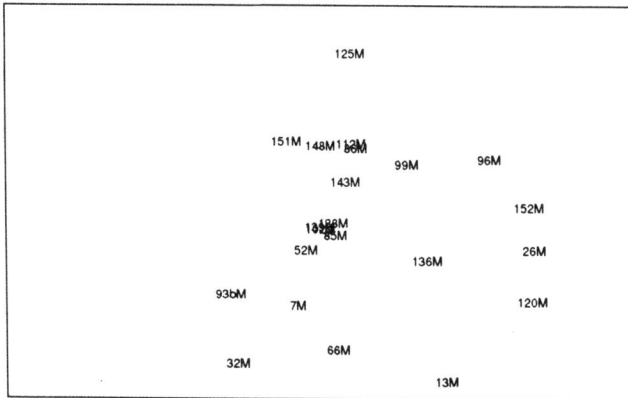

(iii)

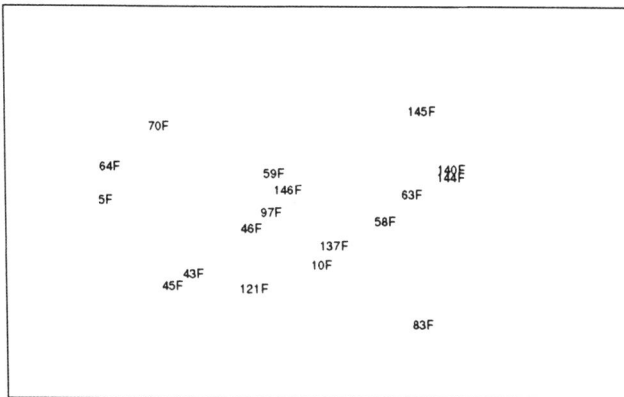

Figures 2.1(ii) and (iii) *Males and females plotted separately*

2.1(iii). The most striking features are that the males tend towards the right of the plot and the females towards the left. The males {99M, 96M, 152M, 26M, 120M, 136M, 66M, 13M} towards the far right tended to have larger mean values for the twelve variables than the rest of the males. The three females {70F, 64F, 5F} to the extreme left of the configuration have mean values much less than those for the other females. Size of skull seems to be the main interpretation horizontally across the configuration.

As a guide to the number of dimensions required for the configuration, the proportion of the variation ($\sum_{i=1}^{p} \lambda_i / \sum_{i=1}^{n} \lambda_i$) explained is 42%, 61%, 78%, 87% and 93% for 1, 2, 3, 4 and 5 dimensional spaces respectively. A three dimensional plot would have been somewhat superior to the two dimensional one shown.

Of course the classical scaling analysis is not the only analysis that could be done on these data. Cluster analysis, discriminant analysis and principal components analysis are all good candidates for such. A principal components analysis could have been carried out on the data – but the resulting configuration of skulls plotted on the first principal component against the second would have been exactly the same as that in Figure 2.1. This is because there is an equivalence between principal components analysis and classical scaling when dissimilarities for classical scaling are chosen to be Euclidean distances. This is explored further in the next section.

2.2.7 Classical scaling and principal components

Suppose $\mathbf{X}$ is a data matrix of dimension $n \times p$. The sample covariance matrix obtained from $\mathbf{X}$ is $\mathbf{S} = (n-1)^{-1} \mathbf{X}^T \mathbf{X}$, where it is assumed that the data have been mean corrected. Principal components are obtained by finding eigenvalues $\{\mu_i : i = 1, \dots, p\}$ and right eigenvectors $\{\boldsymbol{\xi}_i : i = 1, \dots, p\}$ of $\mathbf{S}$, and then the ith principal component is given by $y_i = \boldsymbol{\xi}_i^T \mathbf{x}$ $(i = 1, \dots, p)$ (see for example Chatfield and Collins (1980) or Mardia et $al.$ (1979)).

Suppose, on the other hand, Euclidean distance is used on the data matrix $\mathbf{X}$ to define dissimilarities among the n individuals or objects. The dissimilarities will be given by

$$\delta_{rs}^2 = (\mathbf{x}_r - \mathbf{x}_s)^T (\mathbf{x}_r - \mathbf{x}_s),$$

and hence when these dissimilarities are subjected to classical scaling, $b_{rs} = \mathbf{x}_r^T \mathbf{x}_s$ and $\mathbf{B} = \mathbf{X} \mathbf{X}^T$.

As before let the eigenvalues of $\mathbf{B}$ be λ_i $(i = 1, \dots, n)$ with associated eigenvectors $\mathbf{v}_i$ $(i = 1, \dots, n)$.

It is a well known result that the eigenvectors of $\mathbf{X} \mathbf{X}^T$ are the same as those for $\mathbf{X}^T \mathbf{X}$, together with an extra $n - p$ zero eigenvalues. This is easily shown as

$$\mathbf{X} \mathbf{X}^T \mathbf{v}_i = \lambda_i \mathbf{v}_i.$$

Premultiplying by $\mathbf{X}^T$,

$$(\mathbf{X}^T\mathbf{X})(\mathbf{X}^T\mathbf{v}_i) = \lambda_i(\mathbf{X}^T\mathbf{v}_i).$$

But

$$\mathbf{X}^T\mathbf{X}\boldsymbol{\xi}_i = \mu_i\boldsymbol{\xi}_i,$$

and hence $\mu_i = \lambda_i$ and the eigenvectors are related by $\boldsymbol{\xi}_i = \mathbf{X}^T\mathbf{v}_i$. Thus there is a duality between a principal components analysis and PCO where dissimilarities are given by Euclidean distance. In fact the coordinates obtained in p' dimensions for the n objects by PCO are simply the component scores for the n objects on the first p' principal components. Now $\boldsymbol{\xi}_i^T\boldsymbol{\xi}_i = \mathbf{v}_i^T\mathbf{X}\mathbf{X}^T\mathbf{v}_i = \lambda_i$. Normalizing $\boldsymbol{\xi}_i$, the first p' component scores are given by

$$\mathbf{X}[\lambda_1^{-1}\boldsymbol{\xi}_1, \lambda_2^{-1}\boldsymbol{\xi}_2, \ldots, \lambda_{p'}^{-1}\boldsymbol{\xi}_{p'}] = \mathbf{X}[\lambda_1^{-1}\mathbf{X}^T\mathbf{v}_1, \ldots, \lambda_{p'}^{-1}\mathbf{X}^T\mathbf{v}_{p'}]$$

$$= [\lambda_1^{-\frac{1}{2}}\mathbf{X}\mathbf{X}^T\mathbf{v}_1, \ldots, \lambda_{p'}^{-\frac{1}{2}}\mathbf{X}\mathbf{X}^T\mathbf{v}_{p'}]$$

$$= [\lambda_1^{\frac{1}{2}}\mathbf{v}_1, \ldots, \lambda_{p'}^{\frac{1}{2}}\mathbf{v}_{p'}],$$

which are the coordinates obtained from PCO in p' dimensions.

2.2.8 The additive constant problem

In Chapter 1 various metric and Euclidean properties of dissimilarities were discussed. Here one particular aspect is considered in more detail, that of the additive constant problem. There are two formulations of the problem. The first is simply the problem of finding an appropriate constant to be added to all the dissimilarities, apart from the self-dissimilarities, that makes the matrix $\mathbf{B}$ of Section 2.2 positive semi-definite. This then implies there is a configuration of points in a Euclidean space where the associated distances are equal to the adjusted dissimilarities. This problem has been referred to for many years, Messick and Abelson (1956) being an early reference.

The second formulation is more practically orientated. If dissimilarities are measured on a ratio scale, then there is a sympathy of the dissimilarities to the distances in the Euclidean space used to represent the objects. However if the dissimilarities are measured in an interval scale, where there is no natural origin, then there is not. The additive constant problem can then be stated as the need to estimate the constant c such that $\delta_{rs} + c(1 - \delta_{KR}^{rs})$ may be taken as ratio data, and also possibly to minimize the dimensionality of the Euclidean space required for representing the objects.

For the first formulation, Cailliez (1983) has given an analytic solution. His results are summarized below. The smallest number $c^\star$ has to be found such that the dissimilarities defined by

$$\delta_{rs}^c = \delta_{rs} + c(1 - \delta_{KR}^{rs}) \tag{2.6}$$

have a Euclidean representation for all $c \geq c^\star$, that is which makes the matrix $\mathbf{B}$ positive semi-definite. Let $\mathbf{B}_o(\delta_{rs}^2)$ be the doubly centred matrix based on $\mathbf{A} = [-\frac{1}{2}\delta_{rs}^2]$ for the original dissimilarities. Then substituting δ_{rs}^c for δ_{rs} in (2.6) gives

$$\mathbf{B}_c(\delta_{rs}^2) = \mathbf{B}_0(\delta_{rs}^2) + 2c\mathbf{B}_0(\delta_{rs}) + \tfrac{1}{2}c^2\mathbf{H},$$

noting that $\mathbf{B}_0(\delta_{rs})$ is equivalent to $B_0(\delta_{rs}^2)$ except that the entries are based on δ_{rs} and not δ_{rs}^2.

It is now shown that there exists a constant $c^\star$ such that the dissimilarities $\{\delta_{rs}^c\}$ defined in (2.6) have a Euclidean representation for all $c \geq c^\star$. For $B_c(\delta_{rs}^2)$ to be positive semi-definite it is required that $\mathbf{x}^T\mathbf{B}_c(\delta_{rs}^2)\mathbf{x} \geq 0$ for all $\mathbf{x}$. Now

$$\mathbf{x}^T\mathbf{B}_c(\delta_{rs}^2)\mathbf{x} = \mathbf{x}^T\mathbf{B}_0(\delta_{rs}^2)\mathbf{x} + 2c\mathbf{x}^T\mathbf{B}_0(\delta_{rs})\mathbf{x} + \tfrac{1}{2}c^2\mathbf{x}^T\mathbf{H}\mathbf{x},$$

and so for any $\mathbf{x}$ this gives $\mathbf{x}^T\mathbf{B}_c(\delta_{rs}^2)\mathbf{x}$ as a convex parabola. Therefore, to any $\mathbf{x}$ there corresponds a number $\alpha(\mathbf{x})$ such that $\mathbf{x}^T\mathbf{B}_c(\delta_{rs}^2)\mathbf{x} \geq 0$ if $c \geq \alpha(\mathbf{x})$. Because $\mathbf{B}_0(\delta_{rs}^2)$ is not positive semi-definite, there is at least one $\mathbf{x}$ such that $\mathbf{x}^T\mathbf{B}_0(\delta_{rs}^2)\mathbf{x} < 0$ and for which $\alpha(\mathbf{x})$ will be positive. Hence the number $c^\star = \sup_{\mathbf{x}} \alpha(\mathbf{x}) = \alpha(\mathbf{x}^\star)$ is positive and such that

$$\mathbf{x}^T\mathbf{B}_c(\delta_{rs}^2)\mathbf{x} \geq 0 \quad \text{for all } \mathbf{x} \text{ and all } c \geq c^\star$$

$$\mathbf{x}^{\star T}\mathbf{B}_{c^\star}(\delta_{rs}^2)\mathbf{x}^\star = 0.$$

Hence $\{\delta_{rs}^c\}$ has a Euclidean representation for all $c \geq c^\star$, and also it can be seen that when $c = c^\star$ a space of at most $n - 2$ dimensions is needed since there are now two zero eigenvalues.

Cailliez goes on to find the actual value $c^\star$. He shows that $c^\star$ is given by the largest eigenvalue of the matrix.

$$\begin{bmatrix} \mathbf{0} & 2\mathbf{B}_0(\delta_{rs}^2) \\ -\mathbf{I} & -4\mathbf{B}_0(\delta_{rs}) \end{bmatrix}. \tag{2.7}$$

Cailliez also shows that a negative constant can be added to the original dissimilarities so that

$$\delta_{rs}^c = \mid \delta_{rs} + c(1 - \delta_{KR}^{rs}) \mid,$$

and then a Euclidean representation of $\{\delta_{rs}^c\}$ can be found for all $c < c'$. The value of c' is the smallest eigenvalue of the matrix in (2.7). Going back in time, Messick and Abelson (1956) considered the effect of values of c in (2.6) on the resulting eigenvalues and eigenvectors. They suggested that for a "true" solution, there will be a few large eigenvalues and the rest will be zero or very close to zero. In practice this will not usually be the case and they proposed a method which determined c by setting the mean of the smallest $n - p$ eigenvalues to zero. The largest p eigenvalues are taken as those required to define the Euclidean space. Problems could arise however if large negative eigenvalues occurred. Cooper (1972) included a 'discrepancy' term, η_{rs}, in the new dissimilarities, so that

$$\delta_{rs}^c = \delta_{rs} + c(1 - \delta_{KR}^{rs}) + \eta_{rs},$$

and then c is found for given dimensionality by minimizing $G = \frac{1}{2} \sum_r \sum_s \eta_{rs}^2$. Minimization is done using a Fletcher-Powell routine. The number of dimensions required is then assessed by an index of goodness of fit, FIT:

$$\text{FIT} = 1 - \frac{\sum \eta_{rs}^2}{\sum(\delta_{rs}^c - \delta_{..}^c)^2}.$$

For a perfect solution, FIT $= 1$. To assess the dimension required FIT is plotted against dimension p. The dimension required is that value of p where there is no appreciable improvement in the increase of FIT with increase in p.

Saito (1978) introduced an index of fit, $P(c)$, defined by

$$P(c) = \frac{\sum_{i=1}^p \lambda_i^2(c)}{\sum_{i=1}^n \lambda_i^2(c)},$$

where λ_i is the ith eigenvalue of $B_c(\delta_{rs}^2)$. The constant to be added to the dissimilarities for given P, was then taken as that value which maximizes $P(c)$. Again a gradient method is used for the maximization.

2.3 Robustness

Sibson (1979) studied the effect of perturbing the matrix **B** on the eigenvalues and eigenvectors of $\mathbf{B}$ and hence on the coordinate matrix $\mathbf{X}$. For small ϵ matrix $\mathbf{B}$ is perturbed to $\mathbf{B}(\epsilon)$, where

$$\mathbf{B}(\epsilon) = \mathbf{B} + \epsilon\mathbf{C} + \tfrac{1}{2}\epsilon^2\mathbf{D} + O(\epsilon^3),$$

where $\mathbf{C}, \mathbf{D}$ are symmetric matrices chosen for particular perturbations. The perturbation in $\mathbf{B}$ then causes a perturbation in the eigenvalues λ_i and associated eigenvectors $\mathbf{v}_i$ as follows:

$$\lambda_i(\epsilon) = \lambda_i + \epsilon\mu_i + \tfrac{1}{2}\epsilon^2\nu_i + O(\epsilon^3),$$

$$\mathbf{v}_i(\epsilon) = \mathbf{v}_i + \epsilon\mathbf{f}_i + \tfrac{1}{2}\epsilon^2\mathbf{g}_i + O(\epsilon^3).$$

Sibson shows that

$$\mu_i = \mathbf{v}_i^T\mathbf{C}\mathbf{v}_i, \qquad \mathbf{f}_i = -(\mathbf{B} - \lambda_i\mathbf{I})^T\mathbf{C}\mathbf{v}_i$$

$$\nu_i = \mathbf{v}_i^T(\mathbf{D} - 2\mathbf{C}(\mathbf{B} - \lambda_i\mathbf{I})^+\mathbf{C})\mathbf{v}_i,$$

where $\mathbf{M}^+$ is the matrix $\sum \lambda_i^{-1}\mathbf{v}_i^T\mathbf{v}_i$.

If instead of $\mathbf{B}$ the matrix of squared distances $\mathbf{D}$ is perturbed to a matrix $\mathbf{D}(\epsilon)$

$$\mathbf{D}(\epsilon) = \mathbf{D} + \epsilon\mathbf{F} + O(\epsilon^2),$$

where $\mathbf{F}$ is a symmetric zero diagonal matrix, then the perturbations induced in λ_i and $\mathbf{v}_i$ are

$$\lambda_i(\epsilon) = \lambda_i + \epsilon\mu_i + O(\epsilon^2)$$

$$\mathbf{v}_i(\epsilon) = \mathbf{v}_i + \epsilon\mathbf{f}_i + O(\epsilon^2),$$

where

$$\mu_i = -\tfrac{1}{2}\mathbf{v}_i^T\mathbf{F}\mathbf{v}_i, \qquad \mathbf{f}_i = \tfrac{1}{2}(\mathbf{B} - \lambda_i\mathbf{I})^T\mathbf{F}\mathbf{v}_i + \tfrac{1}{2}(\lambda_i n)^{-1}(\mathbf{1}^T\mathbf{F}\mathbf{v}_i)\mathbf{1}.$$

Matrix $\mathbf{F}$ can be used to investigate various perturbations of the distances $\{d_{rs}\}$. Random errors to the distances can be modelled by assigning a distribution to $\mathbf{F}$ and the effect of these on μ_i and $\mathbf{f}_i$ can be studied.

2.4 Least squares scaling

Until the mid 1970s least squares scaling did not play a major role in multidimensional scaling. Greenacre and Underhill (1982) give a possible explanation:

> It is of interest to note that least squares scaling was the only serious scaling method to be proposed by non-psychologists.

Sibson *et al.* (1981) give some early references to the subject: Spaeth and Guthery (1969) suggest a least squares method in the conclusion of their paper evaluating scaling algorithms; Anderson (1971) likewise proposes the method; Sammon (1969), Chang and Lee (1973) and Bloxom (1978) are other references. Resurgence of least squares MDS occurred when ALSCAL, an alternating least squares method, and SMACOF, a minimization method using a majorizing function were proposed. These are discussed in Chapter 10.

2.4.1 The theory for least squares scaling

Least squares scaling allows a continuous monotonic transformation of dissimilarity $f(\delta_{rs})$ before a configuration is found using least squares. Thus a configuration $\{x_{ri}\}$ is found such that

$$S = \frac{\sum_{r \neq s} w_{rs}(d_{rs} - f(\delta_{rs}))^2}{\sum_{r \neq s} d_{rs}^2}$$

is minimized, where $\{w_{rs}\}$ are appropriately chosen weights.

The distances $\{d_{rs}\}$ do not have to be Euclidean. Minimizing S has to be carried out numerically, and if a gradient method is used then different choices of distance will give rise to different gradient terms.

The weights $\{w_{rs}\}$ can be chosen for specific purposes. For example if $w_{rs} = \delta_{rs}^{-\frac{1}{2}}$ or $w_{rs} = \delta_{rs}^{-1}$ then objects and associated points with small dissimilarities are given more weight than those with large dissimilarities.

The function f has to be decided upon. A straightforward choice is

$$f(\delta_{rs}) = \alpha + \beta\delta_{rs},$$

with α and β estimated in the least squares minimization.

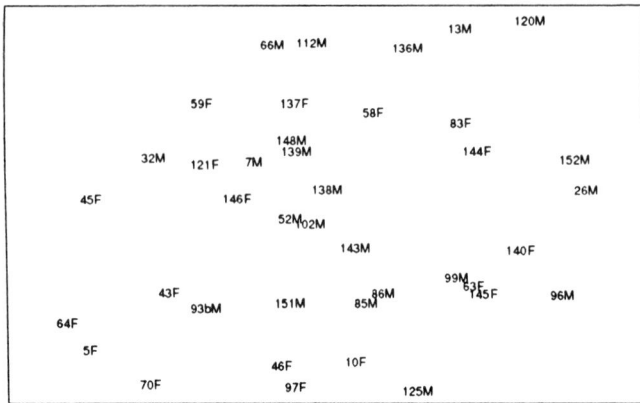

Figure 2.2 *Least squares scaling of the skull data.*

Least squares scaling of the skulls
Figure 2.2 shows the least squares scaling of the skull data from
Section 2.2.6. The special case of the identity function for f was
used, i.e. $f(\delta_{rs}) = \delta_{rs}$. The configuration obtained is in reasonable
agreement with that obtained from classical scaling.

2.4.2 Critchley's intermediate method

Critchley (1978) combines the idea of allowing a transformation
of the dissimilarities in least squares scaling and also the minimization of a target function, with the methods of classical scaling.
Firstly the dissimilarities $\{\delta_{rs}\}$ are transformed using a continuous
parametric function $f(\mu, \delta_{rs})$, where μ is possibly a vector-valued
parameter. For example

$$f(\delta_{rs}) = \delta_{rs}^{\mu} \qquad \mu > 0.$$

Then as for classical scaling

$$\mathbf{B} = \mathbf{HAH}, \quad [\mathbf{A}]_{rs} = -\tfrac{1}{2}f(\delta_{rs}).$$

Let the spectral decomposition of $\mathbf{B}$ be

$$\mathbf{B} = \mathbf{V} \boldsymbol{\Lambda} \mathbf{V}^T,$$

and then μ is estimated by $\hat{\mu}$, the value which minimizes the function

$$T(\mu) = \frac{1}{n^2} \sum_{i=1}^{n} [\lambda_i(\mu)]^2,$$

subject to the constraints

    a) $\lambda_n(\mu) = 0$, so that $\mathbf{B}$ is positive semi-definite, and

    b) $\bar{\lambda} = n^{-1} \sum \lambda_i = 1$, a scale constraint.

See Critchley (1978) for further details.

CHAPTER 3

Nonmetric multidimensional scaling

3.1 Introduction

This chapter presents the underlying theory of nonmetric multidimensional scaling developed in the 1960s. The theory is given for two-way, one-mode data, essentially for dissimilarity data collected on one set of objects.

Suppose there are n objects with dissimilarities $\{\delta_{rs}\}$. The procedure is to find a configuration of n points in a space, which is usually chosen to be Euclidean, so that each object is represented by a point in the space. A configuration is sought so that distances between pairs of points $\{d_{rs}\}$ in the space match "as well as possible" the original dissimilarities $\{\delta_{rs}\}$.

Mathematically, let the objects comprise a set O. Let the dissimilarity, defined on O × O, between objects r and s be δ_{rs} $(r, s \in$ O$)$. Let ϕ be an arbitrary mapping from O onto a set of points X, where X is a subset of the space which is being used to represent the objects. Let the distance between points x_r, x_s in X be given by the real-valued function $d_{x_r x_s}$. Then a disparity, $\hat{d}$, is defined on O × O, which is a measure of how well the distance $d_{\phi(r)\phi(s)}$ "matches" dissimilarity δ_{rs}. The aim is to find a mapping ϕ, for which $d_{\phi(r)\phi(s)}$ is approximately equal to $\hat{d}_{rs}$, and is usually found by means of some loss function. The points in X together with their associated distances will be referred to as a configuration of points.

The choice of dissimilarity measure was discussed in Chapter 1, and it is assumed that dissimilarities $\{\delta_{rs}\}$ have been calculated for the set of objects. The set X is often taken as R^2 and d as Euclidean distance, although others are sometimes used, for example R^3, and the Minkowski metric. Once these are chosen, together with the method for calculating disparities, the nonmetric multidimensional scaling problem becomes one of finding an appropriate algorithm for minimizing a loss function.

A simple example
As a trivial but illustrative example consider the following. Suppose O consists of just three objects labelled $\{1, 2, 3\}$ with dissimilarities given by

$$\delta_{11} = \delta_{22} = \delta_{33} = 0, \delta_{12} = 4, \delta_{13} = 1, \delta_{23} = 3.$$

Let C be a space with just two points $\{a, b\}$, which is used for representing the objects, and so X will be a subset of C. A mapping ϕ then maps each of the three objects in O to one of the two points in C. Thus there must be at least one coincident point. Let the distance function on C be defined as $d_{aa} = d_{bb} = 0, d_{ab} = 1$. Now suppose the disparity function is defined as follows: if the rank order of $\{d_{rs}\}$ is the same as the rank order of $\{\delta_{rs}\}$ then $\hat{d}_{rs} = d_{rs}$, otherwise $\hat{d}_{rs} = 1 - d_{rs}$ for all r, s. Note that the "self-dissimilarities" $\delta_{11}, \delta_{22}, \delta_{33}$ will not be used as is usually the case. The loss function is taken as

$$S = \min_\phi \Big\{ \sum_{r,s} |d_{rs} - \hat{d}_{rs}| \Big\}.$$

There are only eight possible mappings ϕ:

$$\phi_1 : \quad \phi_1(1) = a, \quad \phi_1(2) = a, \quad \phi_1(3) = a$$
$$\phi_2 : \quad \phi_2(1) = a, \quad \phi_2(2) = a, \quad \phi_2(3) = b$$
$$\phi_3 : \quad \phi_3(1) = a, \quad \phi_3(2) = b, \quad \phi_3(3) = a$$
$$\phi_4 : \quad \phi_4(1) = b, \quad \phi_4(2) = a, \quad \phi_4(3) = a$$
$$\phi_5 : \quad \phi_5(1) = a, \quad \phi_5(2) = b, \quad \phi_5(3) = b$$
$$\phi_6 : \quad \phi_6(1) = b, \quad \phi_6(2) = a, \quad \phi_6(3) = b$$
$$\phi_7 : \quad \phi_7(1) = b, \quad \phi_7(2) = b, \quad \phi_7(3) = a$$
$$\phi_8 : \quad \phi_8(1) = b, \quad \phi_8(2) = b, \quad \phi_8(3) = b$$

although only four need to be considered since $\phi_i \equiv \phi_{9-i}$. The rank order of the dissimilarities is $\delta_{13}, \delta_{23}, \delta_{12}$. The possible rank orders of the distances under ϕ_3 for example are d_{13}, d_{12}, d_{23} and d_{13}, d_{23}, d_{12}, giving disparities $\hat{d}_{13} = 1$, $\hat{d}_{12} = \hat{d}_{23} = 0$ and $\hat{d}_{13} = 0$, $\hat{d}_{23} = \hat{d}_{12} = 1$ respectively. The corresponding value of S is 0.0.

The eight values of S under the different mappings are 0.0, 3.0, 0.0, 3.0, 3.0, 0.0, 3.0, and 0.0. So the mappings giving minimum loss are ϕ_1 and ϕ_3 (or ϕ_8 and ϕ_6). The mapping ϕ_1 maps all three objects to a, while ϕ_3 maps objects 1 and 3 to a and 2 to b. In effect the ϕ's carry out a trivial cluster analysis of the three points, ϕ_1 producing only one cluster, and ϕ_3 two clusters.

3.1.1 R^p space and the Minkowski metric

Although nonmetric MDS can be carried out in abstruse spaces the majority of MDS analyses are carried out with X a subset of R^p, and with $p = 2$ in particular. A configuration of points is sought in R^p which represent the original objects, such that the distances between the points $\{d_{rs}\}$ match orderwise, as well as possible, the original dissimilarities $\{\delta_{rs}\}$.

Let the rth point in X have coordinates $\mathbf{x}_r = (x_{r1}, \ldots, x_{rp})^T$.

Let the distance measure for X be the Minkowski metric, and so for points r and s in X,

$$d_{rs} = \left[\sum_{i=1}^{p} |x_{ri} - x_{si}|^\lambda \right]^{1/\lambda} \qquad (\lambda > 0). \qquad (3.1)$$

Define disparities $\{\hat{d}_{rs}\}$, viewed as a function of the distances $\{d_{rs}\}$, by

$$\hat{d}_{rs} = f(d_{rs}),$$

where f is a monotonic function such that

$$\hat{d}_{rs} \leq \hat{d}_{tu} \quad \text{whenever} \quad \delta_{rs} < \delta_{tu} \quad (\text{Condition } C_1).$$

Thus the disparities "preserve" the order of the original dissimilarities but allow possible ties in disparities. Ties in the dissimilarities will be discussed in Section 3.2.5.

Let the loss function be L, where for example

$$L = \left\{ \frac{\sum_{r,s}(d_{rs} - \hat{d}_{rs})^2}{\sum_{r,s} d_{rs}^2} \right\}^{\frac{1}{2}}. \qquad (3.2)$$

Note the original dissimilarities $\{\delta_{rs}\}$ only enter into the loss function by defining an ordering for the disparities $\{\hat{d}_{rs}\}$. The loss function defined above is very commonly used although there are others which will be discussed later. The aim is to find a configuration which attains minimum loss. The loss function can be written in terms of the coordinates $\{x_{ri}\}$ by using equation (3.1) to replace the distances $\{d_{rs}\}$, and can hence be partially differentiated with respect to $\{x_{ri}\}$ in order to seek a minimum. The disparities $\{\hat{d}_{rs}\}$ will usually be a very complicated non-differentiable function of the distances $\{d_{rs}\}$ and hence of the coordinates $\{x_{ri}\}$. This means that the loss function cannot be fully differentiated with respect to the coordinates $\{x_{ri}\}$ when searching for the minimum. Instead various algorithms have been suggested that minimize L with respect to $\{x_{ri}\}$ and also with respect to $\{\hat{d}_{rs}\}$.

Shepard (1962a, 1962b) was the first to produce an algorithm for nonmetric MDS although he did not use loss functions. His method was first to rank and standardize the dissimilarities such that the minimum and maximum dissimilarities were 0 and 1 respectively. Then n points representing the objects are placed at the vertices of a regular simplex in R^{n-1} Euclidean space. Distances $\{d_{rs}\}$ between the n points are then calculated and ranked. The measure of the departure from monotonicity of the distances to the dissimilarities by distance d_{rs} is given by $\delta_{rs} - \delta_{[rs]}$, where $\delta_{[rs]}$ is the dissimilarity of rank equal to the rank of d_{rs}. Shepard's method then moves the points along vectors that will decrease the departure from monotonicity, also stretching larger distances and shrinking smaller distances. The points are repeatedly moved in this manner until adjustments become negligible – however there is no formulation of a proper loss function. After the last iteration, the coordinate system is rotated to principal axes and the first p principal axes are used to give the final configuration in p dimensional space.

It was Kruskal (1964a, 1964b) who improved upon the ideas of Shepard and put nonmetric MDS on a sounder footing by introducing a loss function to be minimized.

3.2 Kruskal's approach

Let the loss function (3.2) be relabelled as S and let

$$S = \sqrt{\frac{S^*}{T^*}}, \qquad (3.3)$$

where $S^* = \sum_{r,s} (d_{rs} - \hat{d}_{rs})^2$, and $T^* = \sum_{r,s} d_{rs}^2$. Note that the summations in the loss function are taken over $1 = r < s = n$ since

$\delta_{sr} = \delta_{rs}$ for all r, s. The loss function is minimized with respect to $\{d_{rs}\}$, i.e. with respect to $\{x_{ri}\}$, the coordinates of the configuration, and also with respect to $\{\hat{d}_{rs}\}$ using isotonic regression.

3.2.1 Minimizing S with respect to the disparities

For convenience let the dissimilarities $\{\delta_{rs}\}$ be relabelled $\{\delta_i : i = 1, \ldots, N\}$ and assume they have been placed in numerical order and that there are no ties. Also relabel the distances $\{d_{rs}\}$ as $\{d_i : i = 1, \ldots, N\}$ where d_i corresponds to the dissimilarity δ_i. To illuminate the proof that follows an example will be employed.

Example
Suppose there are only four objects with dissimilarities

$$\delta_{12} = 2.1, \ \delta_{13} = 3.0, \ \delta_{14} = 2.4, \ \delta_{23} = 1.7, \ \delta_{24} = 3.9, \ \delta_{34} = 3.2$$

and a configuration of points representing the four objects with distances

$$d_{12} = 3.3, \ d_{13} = 4.5, \ d_{14} = 5.7, \ d_{23} = 3.3, \ d_{24} = 4.3, \ d_{34} = 1.3.$$

Then the ordered dissimilarities with the new notation, together with their associated distances, are,

$$\delta_1 = 1.7, \ \delta_2 = 2.1, \ \delta_3 = 2.4, \ \delta_4 = 3.0, \ \delta_5 = 3.2, \ \delta_6 = 3.9$$

$$d_1 = 3.3, \ d_2 = 1.6, \ d_3 = 5.7, \ d_4 = 4.5, \ d_5 = 1.3, \ d_6 = 4.3.$$

Minimization of S is equivalent to minimization of $S' = \sum_i (d_i - \hat{d}_i)^2$, again using the new suffix notation. Let the cumulative sums of $\{d_i\}$ be

$$D_i = \sum_{j=1}^{i} d_j \quad (i = 1, \ldots, N),$$

and consider a plot of D_i against i, giving points $P_0, P_1, \ldots, P_N$ where the origin is labelled P_0. Figure 3.1 shows the plot for the

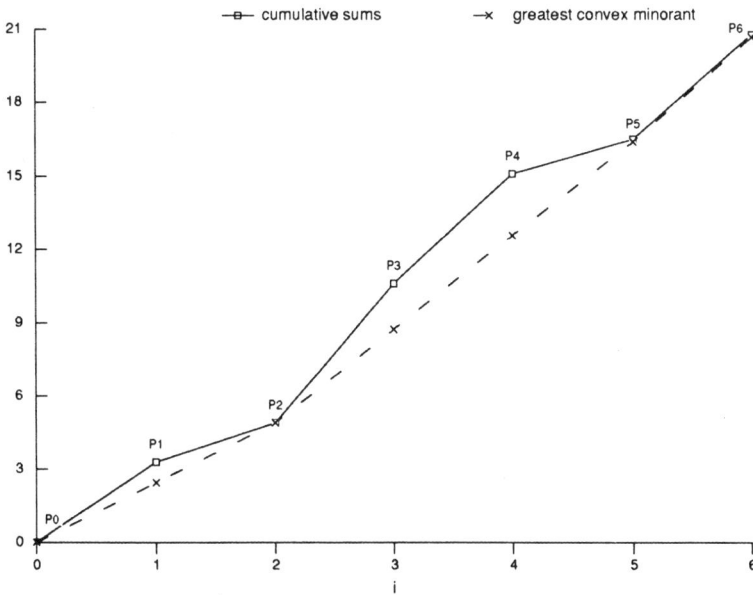

Figure 3.1 *Isotonic regression for the data in the example: solid line – the cumulative sums* $\{D_i\}$, *dashed line – the greatest convex minorant.*

example. Note that the slope of the line joining P_{i-1} and P_i is just d_i.

The greatest convex minorant of the cumulative sums is the graph of the supremum of all convex functions whose graphs lie below the graph of the cumulative sums. (Holding a piece of taut string at P_0 and P_N would give the greatest convex minorant). The greatest convex minorant for the example is also shown in the figure. The $\{\hat{d}_i\}$ which minimizes S' is given by the greatest convex minorant, where $\hat{d}_i$ is the value of the minorant at abscissa i. From Figure 3.1 it is seen that some of the values $\hat{d}_i$ are actually equal to d_i, and obviously $S' = 0$ if $\hat{d}_i = d_i$, for all i. Note that $\hat{d}_i = \hat{D}_i - \hat{D}_{i-1}$ and is the slope of the line. Thus if $\hat{D}_i < D_i$ then $\hat{d}_i = \hat{d}_{i+1}$.

In order to show that this $\{\hat{d}_i\}$ does indeed minimize S', let

$\{d_i^*\}$ be an arbitrary set of real values that satisfy condition C_1. It simply has to be shown that

$$\sum_{i=1}^{N}(d_i - d_i^*)^2 \geq \sum_{i=1}^{N}(d_i - \hat{d}_i)^2.$$

Let

$$D_i^* = \sum_{j=1}^{i} d_i^*, \qquad \hat{D}_i = \sum_{j=1}^{i} \hat{d}_i.$$

Abel's formula, that $\sum_{i=1}^{N} a_i b_i = \sum_{i=1}^{N-1} A_i(b_i - b_{i+1}) + A_N b_N$, where $A_i = \sum_{j=1}^{i} a_j$ are partial sums, will be needed in the following.

Write

$$\sum_{i=1}^{N}(d_i - d_i^*)^2 = \sum_{i=1}^{N}\{(d_i - \hat{d}_i) + (\hat{d}_i - d_i^*)\}^2$$

$$= \sum_{i=1}^{N}(d_i - \hat{d}_i)^2 + \sum_{i=1}^{N}(\hat{d}_i - d_i^*)^2 + 2\sum_{i=1}^{N}(d_i - \hat{d}_i)(\hat{d}_i - d_i^*).$$

Now

$$\sum_{i=1}^{N}(d_i - \hat{d}_i)(\hat{d}_i - d_i^*) = \sum_{i=1}^{N-1}(D_i - \hat{D}_i)(\hat{d}_i - \hat{d}_{i+1})$$

$$- \sum_{i=1}^{N-1}(D_i - \hat{D}_i)(d_i^* - d_{i+1}^*) + (D_N - \hat{D}_N)(\hat{d}_N - d_N^*). \quad (3.4)$$

Now $D_N - \hat{D}_N = 0$ since the last point of the greatest convex minorant and P_N are coincident. Consider $(D_i - \hat{D}_i)(\hat{d}_i - \hat{d}_{i+1})$. If the ith point on the greatest convex minorant is coincident with P_i then $D_i = \hat{D}_i$ and so the term is zero. On the other hand if $\hat{D}_i < D_i$ then $\hat{d}_i = \hat{d}_{i+1}$ and so the term is again zero. Hence since $D_i - \hat{D}_i \geq 0$ and because of the condition C_1, $d_i^* < d_{i+1}^*$ the final term left in (3.4), $-\sum_{i=1}^{N-1}(D_i - \hat{D}_i)(d_i^* - d_{i+1}^*)$, is positive. Hence

$$\sum_{i=1}^{N}(d_i - d_i^*)^2 \geq \sum_{i=1}^{N}(d_i - \hat{d}_i)^2 + \sum_{i=1}^{N}(\hat{d}_i - d_i^*)^2,$$

and so

$$\sum_{i=1}^{N}(d_i - d_i^*)^2 \geq \sum_{i=1}^{N}(d_i - \hat{d}_i)^2.$$

These $\{\hat{d}_{rs}\}$ giving S', and hence S, as a minimum, is the isotonic regression of $\{d_{rs}\}$ (using equal weights) with respect to the simple ordering of $\{\delta_{rs}\}$. Barlow *et al.* (1972) discuss the use of isotonic regression in a variety of situations and illustrate its use in the case of nonmetric MDS. In the MDS literature isotonic regression is referred to as primary monotone least squares regression of $\{d_{rs}\}$ on $\{\delta_{rs}\}$.

So for the illustrative example

$$\hat{d}_1 = \hat{d}_2 = 2.45, \ \hat{d}_3 = \hat{d}_4 = \hat{d}_5 = 3.83, \ \hat{d}_6 = 4.3,$$

noting that $\hat{d}_1, \hat{d}_2$ are the mean of d_1 and d_2; $\hat{d}_3, \hat{d}_4, \hat{d}_5$ are the mean of d_3, d_4 and d_5; $\hat{d}_6$ is equal to d_6. The value of S is 0.14.

3.2.2 A configuration with minimum stress

With $\{\hat{d}_{rs}\}$ defined as the monotone least squares regression of $\{d_{rs}\}$ on $\{\delta_{rs}\}$, S is then termed the stress of the configuration; S^* is called the raw stress. The numerator T^* in the formula for stress is used as a normalizing factor allowing the stress to be dimension free.

A configuration is now sought that minimizes the stress S. Minimization of S is not a particularly easy task. The first step is to place all the coordinates of the points in X in a vector $\mathbf{x} = (x_{11}, \ldots, x_{1p}, \ldots, x_{np})^T$, a vector with np elements. The stress S is then regarded as a function of $\mathbf{x}$, and is minimized with respect to $\mathbf{x}$ in an iterative manner. The method of steepest descent is used, so that if $\mathbf{x}_m$ is the vector of coordinates after the mth iteration

$$\mathbf{x}_{m+1} = \mathbf{x}_m - \frac{\frac{\partial S}{\partial \mathbf{x}} \times sl}{|\frac{\partial S}{\partial \mathbf{x}}|},$$

where sl is the step length discussed later.

Now

$$\frac{\partial S}{\partial x_{ui}} = \frac{1}{2}\sqrt{\frac{T^*}{S^*}}\frac{(T^*\frac{\partial S^*}{\partial x_{ui}} - S^*\frac{\partial T^*}{\partial x_{ui}})}{T^{*2}}$$

$$= \frac{1}{2}S\left(\frac{1}{S^*}\frac{\partial S^*}{\partial x_{ui}} - \frac{1}{T^*}\frac{\partial T^*}{\partial x_{ui}}\right)$$

$$\frac{\partial S^*}{\partial x_{ui}} = 2\sum_{r,s}(d_{rs} - \hat{d}_{rs})\frac{\partial d_{rs}}{\partial x_{ui}}$$

$$\frac{\partial T^*}{\partial x_{ui}} = 2\sum_{r,s}d_{rs}\frac{\partial d_{rs}}{\partial x_{ui}}.$$

For the Minkowski metric

$$\frac{\partial d_{rs}}{\partial x_{ui}} = d_{rs}^{1-\lambda}\sum_{r,s}(x_{ri} - x_{si})^{\lambda-1}(\delta^{ru} - \delta^{su})\mathrm{signum}(x_{ri} - x_{si})$$

and hence

$$\frac{\partial S}{\partial x_{ui}} = S\sum_{r,s}(\delta^{ru} - \delta^{su})\left[\frac{d_{rs} - \hat{d}_{rs}}{S^*} - \frac{d_{rs}}{T^*}\right]$$

$$\times \frac{|x_{ri} - x_{si}|^{\lambda-1}}{d_{rs}^{\lambda-1}}\mathrm{signum}(x_{ri} - x_{si})$$

as given by Kruskal.

A starting configuration giving $\mathbf{x}_0$ needs to be chosen. One possibility is to generate n points according to a Poisson process in a region of R^p. In its simplest form this means simulating each individual coordinate for each point, independently from a uniform distribution on $[0, 1]$. There are several other suggested methods for choosing a starting configuration and these will be discussed in Section 3.6.

Once $\mathbf{x}_0$ has been chosen the method of steepest descent can then be employed to find a configuration with minimum stress using the following algorithm, which is summarized from Kruskal (1964b).

3.2.3 Kruskal's iterative technique

The following summarizes the iterative technique used to find a configuration with minimum stress.

1. Choose an initial configuration.
2. Normalize the configuration to have its centroid at the origin

and unit mean square distance from the origin. This is done since stress is invariant to translation, uniform dilation, and otherwise successive iterations of the procedure might have the configurations continually expanding or wandering around the plane.

3. Find $\{d_{rs}\}$ from the normalized configuration.

4. Fit $\{\hat{d}_{rs}\}$. It was seen that the monotonic least squares regression of $\{d_{rs}\}$ on $\{\delta_{rs}\}$ partitioned $\{\delta_{rs}\}$ into blocks in which the values of $\hat{d}_{rs}$ were constant, and equal to the mean of the corresponding d_{rs} values. In order to find the appropriate partition of $\{\delta_{rs}\}$, first the finest partition is used which has N blocks each containing a single δ_i, using the alternative notation. If this initial partition has $d_1 \leq d_2 \leq \ldots \leq d_N$, then $\hat{d}_i = d_i$ and this partition is the final one. Otherwise two consecutive blocks are amalgamated where $\delta_i > \delta_{i+1}$, and then $\hat{d}_i = \hat{d}_{i+1} = (d_i + d_{i+1})/2$. Blocks are continually amalgamated and new $\hat{d}_i$'s found until the required partition is reached. Full details can be found in Kruskal (1964a) and Barlow *et al.* (1972). The required partition can also be found by considering the graph of the cumulative sums, D_i, and finding the greatest convex minorant. The slope, s_i, of D_i from the origin is D_i/i. The point with the smallest slope must be on the greatest convex minorant. All the points preceding this point are not on the minorant and their slopes can be removed from further consideration. The point with the next smallest slope is then found from those slopes remaining. This point is on the minorant, but the points between the preceding minorant point and this, are not. Their slopes are discarded. This procedure continues until the Nth point is reached. Once the greatest convex minorant has been established it is then an easy task to find $\{\hat{d}_i\}$.

5. Find the gradient $\frac{\partial S}{\partial \mathbf{x}}$. If $|\frac{\partial S}{\partial \mathbf{x}}| < \epsilon$, where ϵ is a preselected very small number, then a configuration with minimum stress has been found and the iterative process can stop. Note that this configuration could be giving a local minimum for the stress, and not the global minimum.

6. Find the new step length sl. Kruskal recommends the ad hoc rule that sl is changed at every step according to

$$sl_{\text{present}} = sl_{\text{previous}} \times (\text{angle factor})$$
$$\times (\text{relaxation factor})$$
$$\times (\text{good luck factor})$$

where

angle factor $= 4.0^{\cos^3 \theta}$,

$\theta =$ angle between the present and previous gradients,

$$\text{relaxation factor} = \frac{1.3}{1 + (5 \text{ step ratio})^5},$$

$$5 \text{ step ratio} = \min \left[1, \left(\frac{\text{present stress}}{\text{stress 5 iterations ago}}\right)\right],$$

$$\text{good luck factor} = \min \left[1, \frac{\text{present stress}}{\text{previous stress}}\right].$$

7. Find the new configuration

$$\mathbf{x}_{n+1} = \mathbf{x}_n - sl\frac{\frac{\partial S}{\partial \mathbf{x}}}{|\frac{\partial S}{\partial \mathbf{x}}|}$$

8. Go to 2.

3.2.4 Nonmetric scaling of breakfast cereals

The 1993 Statistical Graphics Exposition organized by the American Statistical Association contained a data set on breakfast cereals, analyses of which by interested persons could be presented at the Annual Meeting of the Association. Originally observations on eleven variables were collected for seventy-seven different breakfast cereals. For clarity of graphical illustration only those breakfast cereals manufactured by Kellogg are analysed here reducing the number of cereals to twenty-three. The variables measured were: type (hot or cold); number of calories; protein (g); fat (g); sodium (mg); dietry fibre (g); complex carbohydrates (g); sugars (g); display shelf (1,2,3, counting from the floor); potassium (mg); vitamins and minerals (0, 25, or 100, respectively indicating none added; enriched

Table 3.1 *The twenty-three breakfast cereals*

Cereal		Cereal	
All Bran	AllB	Just Right Fruit and Nut	JRFN
All Bran with extra fibre	AllF	Meusliz Crispy Blend	MuCB
Apple Jacks	AppJ	Nut and Honey Crunch	Nut&
Cornflakes	CorF	Nutri Grain Almond Raisin	NGAR
Corn Pops	CorP	Nutri Grain Wheat	NutW
Cracklin Oat Bran	Crac	Product 19	Prod
Crispix	Cris	Raisin Bran	RaBr
Froot Loops	Froo	Raisin Squares	Rais
Frosted Flakes	FroF	Rice Crispies	RiKr
Frosted Mini Wheats	FrMW	Smacks	Smac
Fruitful Bran	FruB	Special K	Spec
Just Right Crunch Nuggets	JRCN		

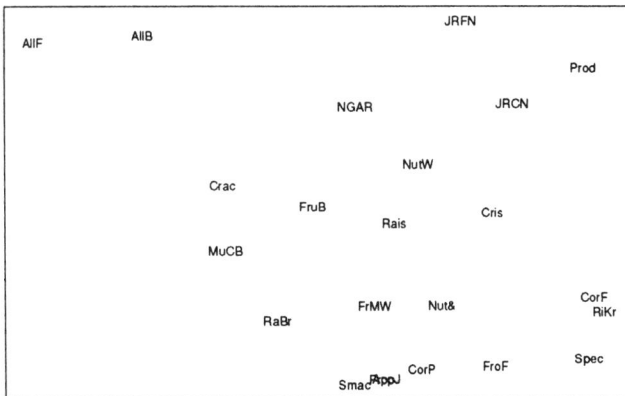

Figure 3.2 *Nonmetric scaling of Kellogg breakfast cereals.*

up to 25% of the recommended daily amount; 100% of the recommended daily amount).

Two dimensional nonmetric scaling was carried out on the Kellogg breakfast cereals, first measuring dissimilarity by Euclidean distance on the variables standardized to have zero mean and unit

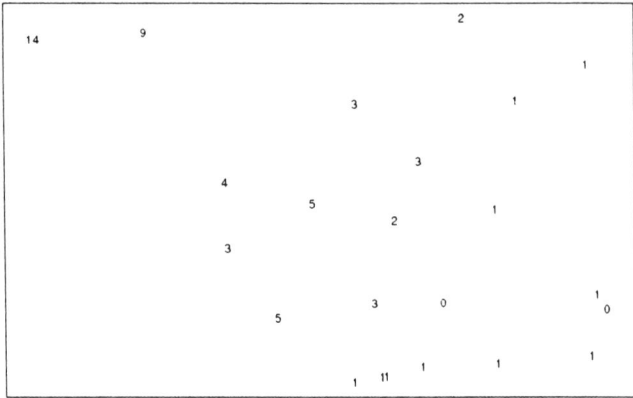

Figure 3.3 *Fibre content of the cereals.*

variance. The stress was 19%. Then using Gower's general dissimilarity coefficient a configuration was found with a 15% stress value. Table 3.1 lists the twenty-three cereals, and Figure 3.2 shows the final configuration. Connoisseurs of breakfast cereals may wish to interpret the configuration. One interesting feature is the spatial pattern of fibre content of the cereals when this is plotted for each cereal at its position in the configuration. Figure 3.3 shows this. Low fibre content is to the lower right of the configuration, high fibre content to the upper left.

Figure 3.4 shows a plot of the dissimilarities $\{\delta_{rs}\}$ against distances $\{d_{rs}\}$ for the configuration together with the isotonic regression of $\{d_{rs}\}$ on $\{\delta_{rs}\}$, i.e. the disparities. This is known as the Shepard diagram and is useful in assessing the fit. Note that the Shepard diagram is usually plotted with the axes of Figure 3.4 reversed in accordance with usual regression practice. Preference depends on how the figure is to be viewed, either the isotonic regression of $\{d_{rs}\}$ on $\{\delta_{rs}\}$, or disparities plotted against $\{d_{rs}\}$.

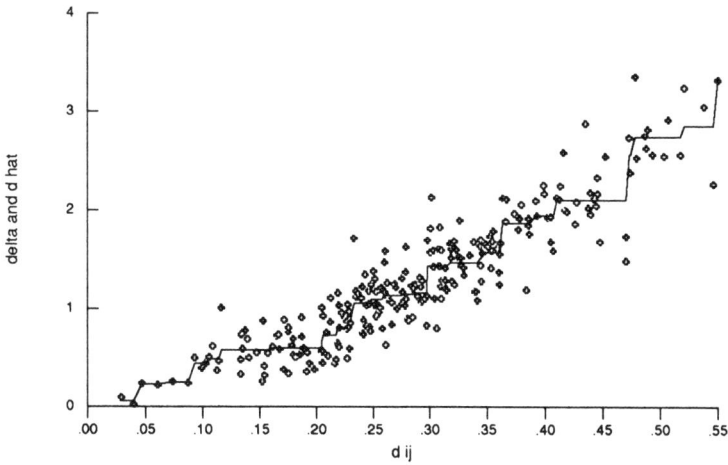

Figure 3.4 *Shepard diagram for the breakfast cereal data.*

3.2.5 STRESS1/2, monotonicity, ties and missing data

The stress function (3.3) used by Kruskal is often referred to as STRESS1. An alternative stress function is sometimes employed in nonmetric MDS, given by

$$
S = \left\{ \frac{\sum_{r,s}(d_{rs} - \hat{d}_{rs})^2}{\sum_{r,s}(d_{rs} - d_{..})^2} \right\}^{\frac{1}{2}},
$$

where $d_{..}$ is the mean of the distances $\{d_{rs}\}$ over $1 \leq r < s \leq n$. This is referred to as STRESS2. Only the normalizing factor differs in the two definitions of stress.

Recall condition C_1 that

$$
\hat{d}_{rs} \leq \hat{d}_{tu} \quad \text{whenever} \quad \delta_{rs} < \delta_{tu}.
$$

This is referred to as the weak monotonicity condition and the fitted $\{\hat{d}_{rs}\}$ are weakly monotone with the data. This condition can be replaced by condition C_2 that

$$
\hat{d}_{rs} < \hat{d}_{tu} \quad \text{whenever} \quad \delta_{rs} < \delta_{tu} \quad \text{(Condition } C_2).
$$

This is the strong monotonicity condition and the fitted $\{\hat{d}_{rs}\}$ are strongly monotone with the data. This latter case will give larger stress values since more restriction is placed on the configuration.

There are two ways that ties in the dissimilarities can be treated. The primary approach is:

If $\delta_{rs} = \delta_{tu}$ then $\hat{d}_{rs}$ is not necessarily equal to $\hat{d}_{tu}$.

The secondary approach is:

If $\delta_{rs} = \delta_{tu}$ then $\hat{d}_{rs} = \hat{d}_{tu}$.

The secondary approach is very restrictive and has been shown by many authors, for example Kendall (1971) and Lingoes and Roskam (1973), to be less satisfactory than the primary approach. Kendall (1977), in an appendix to Rivett (1977), introduces a tertiary approach to ties which is a hybrid between the primary and secondary approaches.

One desirable aspect of nonmetric MDS is that if some of the dissimilarities are missing then they are simply left out of the formula for stress, and the fitting algorithm proceeds without them. Consequences of missing data are discussed in Section 3.4 where further results about stress are reported.

3.3 The Guttman approach

Guttman (1968) took a different approach to Kruskal (1964a,b) in setting up nonmetric MDS. He defined a loss function called the coefficient of alienation which was basically equivalent to the stress function of Kruskal, but which led to a different algorithm for minimization. His approach will only be described briefly.

Let the rank ordered dissimilarities $\{\delta_{rs}\}$ be placed in a vector δ with elements δ_r $(r = 1, \ldots, N)$. Let the distances $\{d_{rs}\}$ from a configuration be placed in a vector $\mathbf{d}$ in order corresponding to $\{\delta_r\}$. Let $\mathbf{E}$ be an $N \times N$ permutation matrix which places the elements of $\mathbf{d}$ into ascending order. Disparities are then defined by the rank-image $\mathbf{d}^*$ of $\mathbf{d}$, given by

$$\mathbf{d}^* = \mathbf{E}\mathbf{d}.$$

The coefficient of continuity, μ, for the configuration is given by

$$\mu = \sqrt{\frac{(\sum d_r d_r^*)^2}{\sum d_r^2 \sum d_r^{*2}}}$$

which has the value unity for a perfect fit. In order to find a best fitting configuration the coefficient of alienation, K, given by

$$K = \sqrt{1 - \mu^2}$$

is minimized using the method of steepest descent.

Example
Suppose there are only three objects, with dissimilarities

$$\delta_{12} = 4, \ \delta_{13} = 1, \ \delta_{23} = 3,$$

with "self-dissimilarities" zero. Let a particular configuration have distances between its points

$$d_{12} = 2, \ d_{13} = 4, \ d_{23} = 5.$$

Then in the single suffix notation, and ordering the dissimilarities,

$$\delta_i \quad : \quad 1, \quad 3, \quad 4$$
$$d_i \quad : \quad 4, \quad 5, \quad 2.$$

The permutation matrix $\mathbf{E}$ is

$$\mathbf{E} = \begin{bmatrix} 0 & 0 & 1 \\ 1 & 0 & 0 \\ 0 & 1 & 0 \end{bmatrix}$$

giving $\mathbf{d}^* = (2, 4, 5)^T$.

The coefficient of continuity, μ is then 0.84, and the coefficient of alienation K is 0.54.

Guttman's paper is much more detailed than this simple exposition, dealing with strong and weak monotonicity and ties in the data. It can be shown that minimizing K is equivalent to minimizing stress S. Guttman and Lingoes produced a series of computer programs for nonmetric MDS based on the Guttman approach, and these are included in their SSA-I (smallest space analysis) series of programs.

They use two main strategies for minimization. Their single phase G-L algorithm minimizes

$$\phi^* = \sum (d_{rs} - d_{rs}^*)^2,$$

using the method of steepest descent. For brevity the various derivatives similar to those for the Kruskall algorithm are not written down here but can be found in Lingoes and Roskam (1973), or

Davies and Coxon (1983). Their double-phase G-L algorithm first minimizes ϕ^* with respect to $\{d_{rs}\}$ as its first phase, i.e. finds the configuration that best fits the current values $\{d_{rs}^*\}$. The second phase then finds new values $\{d_{rs}^*\}$ which best fit the new configuration.

3.4 A further look at stress

Several authors have studied stress in more detail. We report on some of their results.

Differentiability of stress
Because of the complicated nature of stress through the involvement of least squares monotone regression, continuity and differentiability of stress and its gradient could cause concern when seeking a minimum. However Kruskal (1971) shows that $\sum(d_i - \hat{d}_i)^2$ has gradient vector with ith element $2(d_i - \hat{d}_i)$ and that the gradient exists and is continuous everywhere.

De Leeuw (1977a) noted that the Euclidean distance between two points, $d_{rs} = \{\sum_i(x_{ri} - x_{si})^2\}^{\frac{1}{2}}$ is not differentiable in a configuration if points x_r and x_s are coincident. De Leeuw (1977b) shows that gradient algorithms can be modified to cope with the problem. De Leeuw (1984) shows that when stress is minimized coincident points cannot occur.

Limits for stress
The minimum possible value of Kruskall's stress is zero, implying a perfect fit. However a zero stress value can imply that the final configuration is highly clustered with a few tight clusters of points.

De Leeuw and Stoop (1984) give upper bounds for stress. Let STRESS1 be denoted by $S(n,p)$, where the number of points, n, and the dimension, p, of the configuration are fixed. They show that

$$S(n,p) \leq \kappa(n,p),$$

where

$$\kappa(n,p) = \min_{x_{ri}} \left\{ \sqrt{\frac{\sum_{r,s}(d_{rs} - d..)^2}{\sum_{r,s} d_{rs}^2}} \right\},$$

with $d..$ the usual mean of $\{d_{rs}\}$ over r, s.

This result is easily seen as

$$\sum_{r,s}(d_{rs} - \hat{d}_{rs})^2 \leq \sum_{r,s}(d_{rs} - d_{..})^2,$$

since disparities defined as $\hat{d}_{rs} = d_{..}$ for all r, s satisfy the monotonicity requirement, but obviously do not minimize the stress or raw stress over the disparities, since this is achieved by $\{\hat{d}_{rs}\}$, the isotonic regression of $\{d_{rs}\}$ on $\{\delta_{rs}\}$. Dividing by $\sum_{r,s} d_{rs}^2$, taking square roots and minimizing over the configuration $\{x_{ri}\}$ proves the result.

De Leeuw and Stoop (1984) then go on to show that

$$\kappa(n,p) \leq \kappa(n,1) = \left(\frac{n-2}{3n}\right)^{1/2} \leq \frac{1}{\sqrt{3}} = 0.5774.$$

It is easily seen that $\kappa(n,p) \leq \kappa(n,1)$ since in minimizing

$$\sqrt{\frac{\sum_{r,s}(d_{rs} - d_{..})^2}{\sum_{r,s} d_{rs}^2}} \tag{3.5}$$

over the p dimensional configuration $\{x_{ri}\}$, it is always possible to take the configuraton as the projection onto the single axis used for $\kappa(n,1)$ (or any other subspace of dimension less than p).

To calculate $\kappa(n,1)$, assume without loss of generality $\sum x_r = 0$, and $\sum x_r^2 = 1$. Then

$$\sum_{r,s} d_{rs}^2 = \frac{1}{2}\sum_{r=1}^{n}\sum_{s=1}^{n} d_{rs}^2 = \frac{1}{2}\sum_{r=1}^{n}\sum_{s=1}^{n}(x_r - x_s)^2 = n,$$

and hence minimizing (3.5) is equivalent to minimizing

$$\sqrt{1 - \frac{(n-1)}{2}d_{..}^2},$$

which in turn is equivalent to maximizing $d_{..}$.

Reordering the points $\{x_r\}$ such that $x_1 \leq x_2 \leq \ldots \leq x_n$, it is seen that $d_{..}$ is given by

$$d_{..} = \frac{2}{n(n-1)}\sum_{r=1}^{n}(2r - n - 1)x_r.$$

Now $d_{..}$ is maximized when $\{x_r\}$ are equally spaced along the axis. Let $x_r = a + br$. Hence

$$\sum (a + br) = na + \tfrac{1}{2}n(n + 1)b = 0,$$

and

$$\sum (a + br) = na^2 + n(n + 1)ab + \tfrac{1}{3}n(n + 1)(2n + 1)b^2 = 1.$$

Solving gives

$$x_r = \sqrt{\frac{12}{n(n^2 - 1)}} \left\{ r - \frac{(n + 1)}{2} \right\}.$$

Hence $\kappa(n, 1) = (\frac{n-2}{3n})^{\frac{1}{2}}$ after some algebra, and it is easily seen that $\kappa(n, 1)$ tends to $1/\sqrt{3}$ as n tends to infinity.

De Leeuw and Stoop also give an upper bound for $\kappa(n, 2)$ and show

$$\kappa(n, 2) \leq \kappa^*(n, 2) = \left\{ 1 - \frac{2 \cot^2(\pi/2n)}{n(n - 1)} \right\}^{\frac{1}{2}} \leq \left\{ 1 - \frac{8}{\pi^2} \right\}^{\frac{1}{2}} = 0.4352.$$

The value of $\kappa^*(n, 2)$ is the value of (3.5) if $\{x_{ri}\}$ consists of n equally spaced points on a circle. Note that $\kappa(n, 2)$ is not necessarily equal to $\kappa^*(n, 2)$.

For STRESS2 it can easily be seen from (3.5) that the upper limit for STRESS2 is unity.

3.4.1 Interpretation of stress

Since Kruskal's 1964 papers in *Psychometrika* there have been many investigations of stress using Monte Carlo methods. Various findings are reported below.

Stenson and Knoll (1969) suggested that in order to assess dimensionality and fit of the final configuration, stress values for a set of dissimilarity data should be compared with those obtained using random permutations of the first $\binom{n}{2}$ integers as dissimilarities. They used $n=10(10)60$, $p=1(1)10$ in a Monte Carlo study, using three random permutations for each combination of n and p. They plotted mean stress against dimension p, for a fixed number of objects n. They managed with only three random permutations since the variability of stress was small. Spence and Ogilvie (1973) carried out a more thorough Monte Carlo study using fifteen replications for each n, p combination. De Leeuw and Stoop

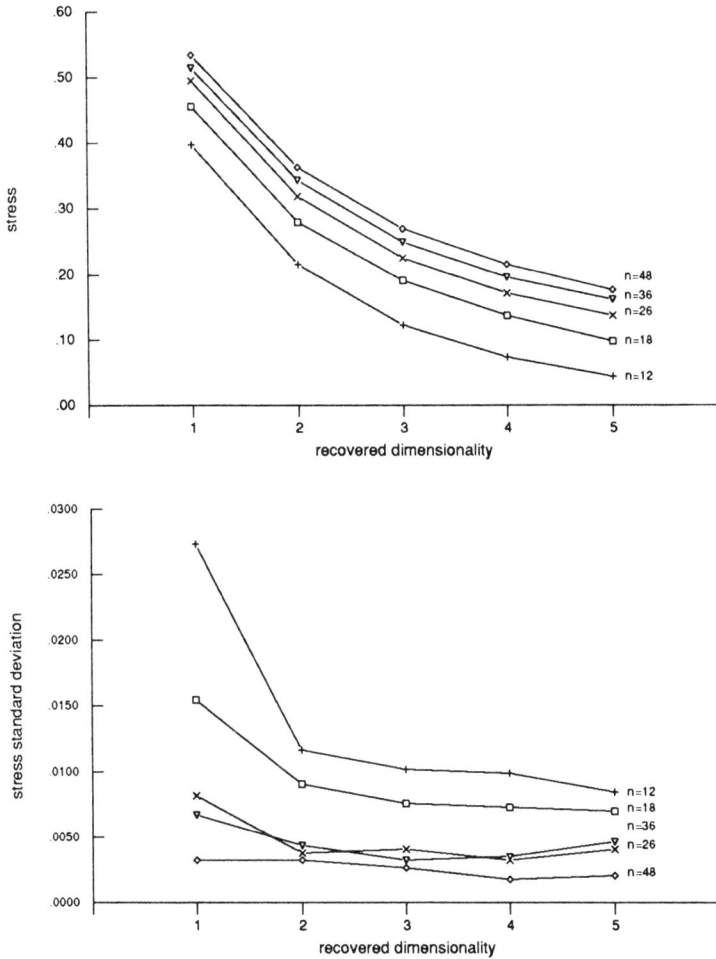

Figure 3.5 *Mean stress and standard deviation obtained from random rankings (Spence and Ogilvie, 1973).*

(1984) carried out a similar exercise using one hundred random rankings. Spence and Ogilvie's results for mean and standard deviation of stress are shown in Figure 3.5. The mean stress is useful since it gives a guide as to whether the stress obtained in a study is too large or not for a reasonably fitting final configuration. Levine (1978) carried out a similar exercise using Kruskal's STRESS2 in place of STRESS1.

Klahr (1969) generated $\binom{n}{2}$ dissimilarities $\{\delta_{rs}\}$ independently from a uniform distribution on [0,1], choosing n as 6,7,8,10,12 and 16, $p=1(1)5$, and subjected them to nonmetric MDS. This was done one hundred times for smaller values of n and fifty times for larger values, for each value of p. The sample distribution function for stress was plotted, as well as a plot of mean stress. Klahr noted that it was often possible to obtain a well fitting final configuration of points for small values of n even when the dissimilarities were randomly generated in this manner.

Spence (1970) generated configurations of points according to a p dimensional Poisson process within a unit hypersphere. To each individual coordinate an independent normally distributed random "error" was added. Dissimilarities were then taken as the Euclidean distances between the pairs of points and these were subjected to nonmetric MDS. The stress values obtained in attempting to retrieve the original configurations were used to compare the frequency with which the three MDS programs, TORSCA, MDSCAL and SSA-1 got stuck in local minimum solutions. Spence (1972) went on to compare the three programs in more depth. Wagenaar and Padmos (1971) carried out a simulation study of stress in a similar manner to that of Spence using a realization of a p dimensional Poisson process. However their dissimilarities were taken as the Euclidean distance between pairs of points in the configuration, together with multiplicative error, introduced by multiplying the distances by an independent random number generated from a normal distribution. They used their stress results in a method to assess the required dimensionality of the configuration. This method is explained in Section 3.5.

Sherman (1972) used the p dimensional Poisson process to generate configurations, choosing $n=6,8,10,15,30$, $p=1,2,3$. An independent normal error was added to each coordinate and dissimilarities were generated using the Minkowski metric with $\lambda = 1, 2, 3$. Sherman used analysis of variance to investigate the factors most affecting nonmetric MDS results, and concluded with basic common sense suggestions, such as that hypothesized structure should be of low dimension, measurement errors should be minimized, and various dimensions should be tried for the configuration with varying λ in the Minkowski metric.

Sibson et al. (1981) consider more sophisticated models for producing dissimilarities from distances obtained from a configuration

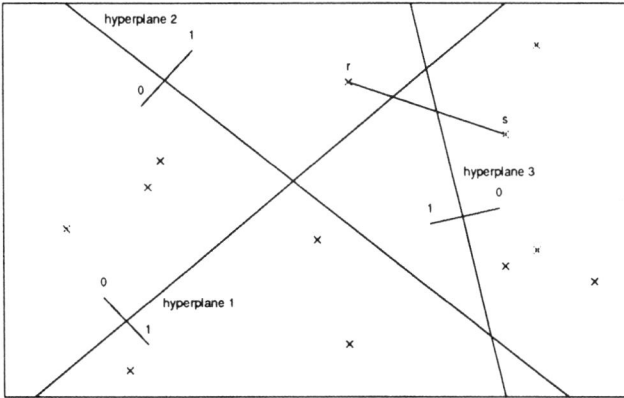

Figure 3.6 *Points r and s in two dimensional Euclidean space with three random hyperplanes*

of points before attempting to recover the orginal configuration using nonmetric MDS. Their first model is based on binary data and the Hamming distance. Suppose there are k binary variables measured on each of the objects. Then the Hamming distance between objects r and s is simply the number of variables in which the two objects differ, and thus is very closely related to the dissimilarity coefficients of Chapter 1. Consider two points r and s in a p dimensional Euclidean space together with a Poisson hyperplane process where random hyperplanes cut the space into two half spaces. The two half spaces are denoted zero and one arbitrarily. Figure 3.6 shows Euclidean space with $p=2$, and three hyperplanes, with associated zeros and ones allocated.

The binary data associated with the point r is $(0,1,1)$ and that with point s $(1,0,1)$. The Hamming distance is 2. In general the Hamming distance is equal to the number of hyperplanes crossed in going from one point to another, and can be randomly generated by randomly locating these hyperplanes. From Figure 3.6 the number of hyperplanes crossing the line between points r and s is two, in agreement with the data. The number of hyperplanes crossing the line between point r and point s follows a Poisson

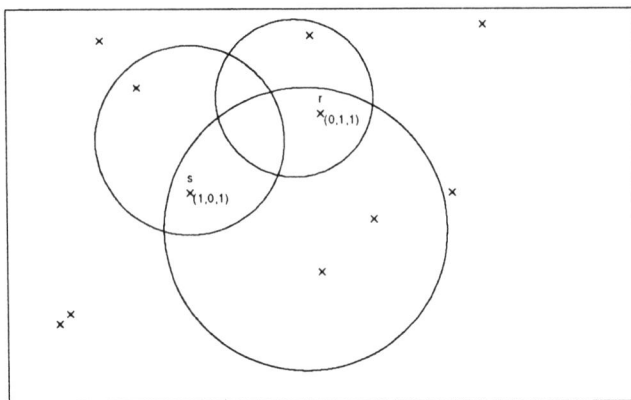

Figure 3.7 *Model for binary data and the Jaccard coefficients*

distribution with parameter equal to λd_{rs} where λ is the intensity of the process. Conditioned upon the total number of hyperplanes the distribution of the Hamming distance is a binomial distribution. Thus for their first model, Sibson *et al.* generate a realization from a p dimensional Poisson process and then split the space with Poisson hyperplanes. The dissimilarity between points r and s is then taken as the Hamming distance between these points.

Their second model is similar to the first model but has dependence between points removed. The dissimilarity δ_{rs} is taken as a random number from a Poisson distribution with parameter λd_{rs}.

Sibson *et al.*'s third model generates random Jaccard coefficients. Each binary variable is considered to measure presence (1), or absence (0). A realization of a p dimensional Poisson process again starts off the model. Then for each variable a p dimensional hypersphere is generated with radius randomly chosen from some distribution, and centre a point in another realization of a Poisson process. Inside each hypersphere the variable assumes the value unity, and outside the hypersphere the value zero. For example Figure 3.7 shows ten points and three variables for a two dimensional space. The point r has binary data associated with it $(0,1,1)$

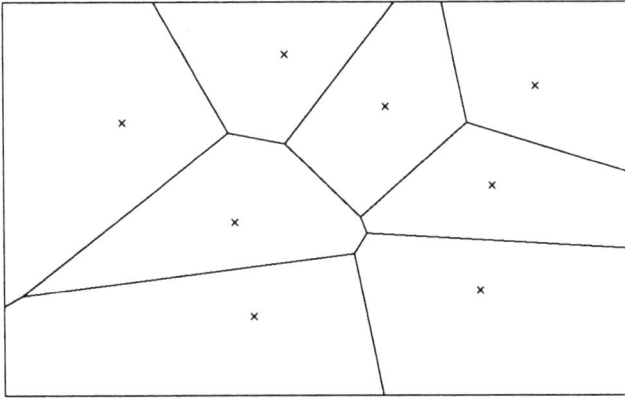

Figure 3.8 *Dirichlet tesselation model*

and point s has $(1,0,1)$, and hence $\delta_{rs} = 1/3$. If r and s were both to lie outside all the spheres, the dissimilarity would be unity.

For their fourth model points are again generated from a p dimensional Poisson process but then Dirichelet tesselations are found; see Green and Sibson (1978). Figure 3.8 shows a two dimensional example. Dirichelet tesselations are found for each point r, a surrounding polygon where all points of the space within the polygon are closer to r than any other. The Wilkinson metric for points r and s is then the minimum number of boundaries crossed to get from one point to the other. The dissimilarity δ_{rs} is then taken as this Wilkinson distance.

Sibson *et al.* use the Procrustes statistic (see Chapter 5) to compare recovered configurations using nonmetric MDS with the original configurations. They use three scaling methods: classical MDS, Kruskal's nonmetric MDS, and least squares scaling. Among their conclusions they maintain that classical MDS compares well with nonmetric MDS for most "Euclidean-like" models, but not for "non-Euclidean-like" models. Least squares scaling is slightly superior to nonmetric MDS for the Euclidean-like models but inferior for the Jaccard coefficient model. Nonmetric MDS is never significantly worse than the other methods if it is given a reasonable starting configuration.

The models in Sibson *et al.* (1981) have been described in detail here even though their investigation does not deal directly with stress, because they use models which generate dissimilarities of an interesting nature. Most other Monte Carlo investigations mundanely have additive or multiplicative noise applied to coordinates or distances in order to produce dissimilarities.

All the stress studies have shown that stress decreases with increase of dimension p, increases with the number of points n, and that there is not a simple relationship between stress, n and p. By using a different model for error Cox and Cox (1990), and Cox and Cox (1992) found a simple relationship between stress, n and p. Up till then most Monte Carlo investigations started with a configuration of points generated according to a Poisson process. Cox and Cox considered configurations covering a wide range of spatial patterns; see for example Ripley (1981) or Diggle (1983) for full discussion of spatial stochastic models. At one extreme was the highly regular process of a rectangular grid. For a pattern with less regularity, this rectangular grid had its points independently radially displaced by an amount (R, Θ) where R has a Rayleigh distribution (pdf $r\sigma^{-2} \exp(-\frac{1}{2}r^2/\sigma^2)$) and Θ a uniform distribution on $[0, 2\pi]$. The further the average displacement the less regular the process becomes, and in the limit the process tends to a Poisson process. At the other extreme points were generated according to a Poisson cluster process. Here cluster centres are generated according to a Poisson process and then a fixed number of cluster members are positioned at radial points (R, Θ) from the cluster centre, where R, Θ are as above. As the points in a cluster are moved further and further away from their cluster centres, the process tends towards a Poisson process again. Thus a very wide range of spatial patterns were considered ranging from extreme regularity on one hand, through complete spatial randomness (i.e. the Poisson process) to extreme aggregation on the other. Figure 3.9 shows realizations for the three models.

For each configuration generated, dissimilarities were defined as

$$\delta_{rs} = d_{rs}(1 - \epsilon_{rs}),$$

where d_{rs} is the usual Euclidean distance and $\{\epsilon_{rs}\}$ are independent uniformly distributed random variables on the interval $[0, l]$.

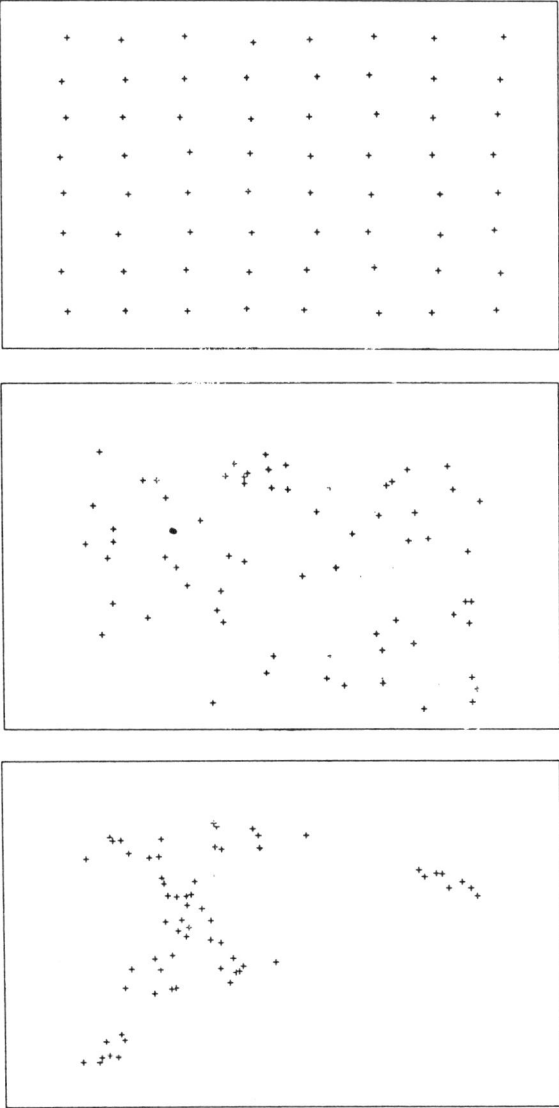

Figure 3.9 *Realizations of the three models considered by Cox and Cox: regular process; Poisson process; Poisson cluster process.*

The value of l can be considered as the noise level. Several values of n and l were chosen. The number of dimensions, p, was chosen

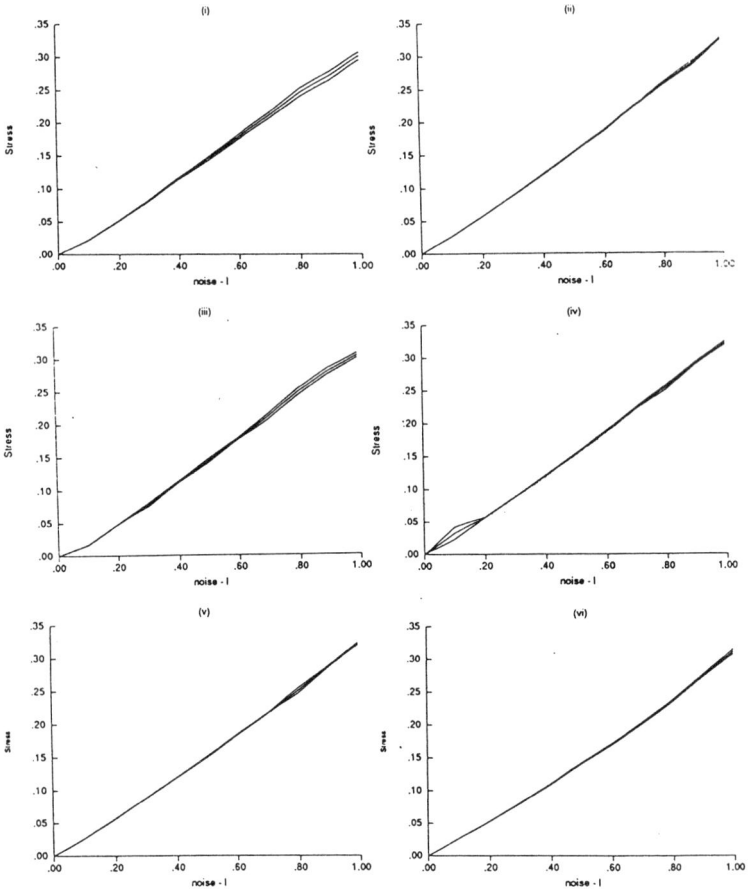

Figure 3.10 *Stress plotted against noise l for various spatial models*

as only $p = 2$ in Cox and Cox (1990), but results were extended for other values of p in Cox and Cox (1992). Nonmetric MDS was used on the dissimilarities generated for each configuration and stress recorded. Each model was replicated ten times and average stress found. In keeping with other authors' results, the variability in stress for fixed n, p and l was extremely small. For two dimensional initial configurations, together with a derived configuration also in two dimensions, the following results were observed.

Figure 3.10 shows average stress plotted against the "noise level"

l for (i) a Poisson process, $n = 36$; (ii) a Poisson process, $n = 64$; (iii) a rectangular grid, $n = 36$; (iv) a regular process, $n = 64$, $\sigma^2 = 0.25$; (v) a Poisson cluster process, 16 clusters of size 4, $\sigma^2 = 0.1$, and (vi) a Poisson cluster process, 4 clusters of size 16, $\sigma^2 = 0.1$.

The remarkable result is that with this noise model, stress is proportional to noise level l ($l \doteq 3 \times$ stress) whatever the value of n and for all reasonable spatial patterns of points (i.e. ones which are not comprised of a few very tight clusters). This means that if the model is reasonable then stress levels for different sets of dissimilarities can be directly compared for any differing number of objects and for any spatial pattern formed by the final configuration of points. For dimensions other than two, similar results were found by Cox and Cox (1992), but with not such a strong linear relationship between stress and l. It should be noted that these results only hold when the dimension of the configuration derived by MDS is the same as that of the original configuration used to generate the dissimilarities.

3.5 How many dimensions?

For illustrative purposes the obvious preferred number of dimensions to be chosen for nonmetric MDS is two. Configurations in three dimensions can be illustrated using three dimensional plotting procedures from various statistical packages, such as SAS, SOLO and STATISTICA. However a less well fitting configuration in two dimensions may be preferable to one in several dimensions where only projections of points can be graphically displayed.

To choose an appropriate number of dimensions, Kruskal (1964a) suggests that several values of p, the number of dimensions, are tried and the stress of the final configuration plotted against p. Stress always decreases as p increases. Kruskal suggests that p is chosen where the "legendary statistical elbow" is seen in the graph. For the breakfast cereal data of Section 3.2.4 this was done and results are shown in Figure 3.11.

The "elbow" appears to be at $p = 4$. However it has been noted that often there is no sharp flattening of stress in these diagrams and that an elbow is hard to discern.

Wagenaar and Padmos (1971) suggested the following method for choosing the appropriate number of dimensions. Dissimilarities are subjected to nonmetric MDS in 1, 2, 3,... dimensions, and the

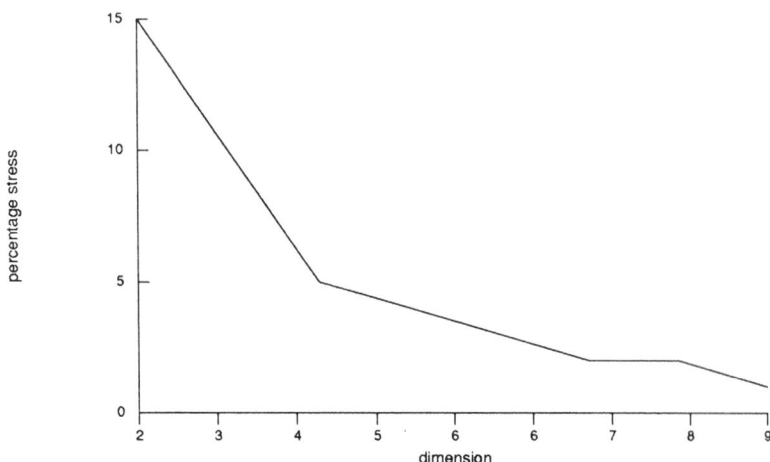

Figure 3.11 *Stress plotted against dimension for the breakfast cereal data*

values of stress noted in each case, say S_1, S_2, S_3,... . These are then compared to the stress results from Monte Carlo simulations where dissimilarities are generated from distances in spatial configurations of points, together with random noise. The level of noise, σ_1 needed in one dimension to give stress S_1 is noted. Then for two dimensions, S_2 is compared with S_2^E, the "expected stress" for two dimensions with noise level σ_1. If S_2 is significantly less than S_2^E then the second dimension is definitely needed. The noise level σ_2, needed to produce a stress level S_2 in two dimensions, is found, and then the "expected stress" S_3^E, for three dimensions with this noise level, σ_2. The stress S_3 for three dimensions is then compared with S_3^E. This process continues until stress is comparable to the expected stress, implying that there is no gain to be made in increasing the number of dimensions beyond that point.

3.6 Starting configurations

One possibility for a starting configuration for nonmetric MDS algorithms is simply to use an arbitrary one. Points can be placed at the vertices of a regular p dimensional lattice for instance, or could be generated as a realization of a p dimensional Poisson process.

This latter case simply requires all coordinates to be independently generated from a uniform distribution on $[-1, 1]$ say, and the configuration is then normalized in the usual manner to have centroid at the origin and mean squared distance of the points from the origin, unity. It is always recommended that several different starting configurations are tried in order to avoid local minimum solutions.

If metric MDS is used on the data initially, the resulting configuration can be used as a starting configuration for nonmetric MDS.

Guttman (1968) and Lingoes and Roskam (1973) suggested the following for finding a starting configuration. Let matrix $\mathbf{C}$ be defined by $[\mathbf{C}]_{rs} = c_{rs}$, where

$$c_{rs} = 1 + \sum_i \rho_{ri}/N \quad (r = s)$$

$$= 1 - \rho_{rs}/N \quad (r \neq s),$$

where N is the total number of dissimilarities $\{\delta_{rs}\}$, and ρ_{rs} is the rank of δ_{rs} in the numerical ordering of $\{\delta_{rs}\}$. The principal components of $\mathbf{C}$ are found and the initial configuration is given by the eigenvectors of the first p principal components, but ignoring the one with constant eigenvector.

3.7 Interesting axes in the configuration

A simple method for finding meaningful directions or axes within the final configuration is to use multiple linear regression. The method is explained in Kruskal and Wish (1978). An axis is found for a variable, y, related to the objects, or even one of the original variables used in defining the dissimilarities. This variable is taken as the dependent variable. The independent variables are the coordinates of the points in the final configuration.

The regression model is then

$$\mathbf{y} = \mathbf{X}\boldsymbol{\beta} + \boldsymbol{\epsilon},$$

where $\mathbf{y}$ is the vector of observations $\{y_i\}$ $(i = 1, \ldots, n)$, $\mathbf{X}$ is the $n \times (p + 1)$ matrix consisting of a column of ones followed by the coordinates of the points in the final configuration, $\boldsymbol{\beta}$ is the parameter vector, and $\boldsymbol{\epsilon}$ the "error" vector.

The least squares estimate of $\boldsymbol{\beta}$ is given by

$$\hat{\boldsymbol{\beta}} = (\mathbf{X}^T\mathbf{X})^{-1}\mathbf{X}^T\mathbf{y}.$$

(i)

(ii)

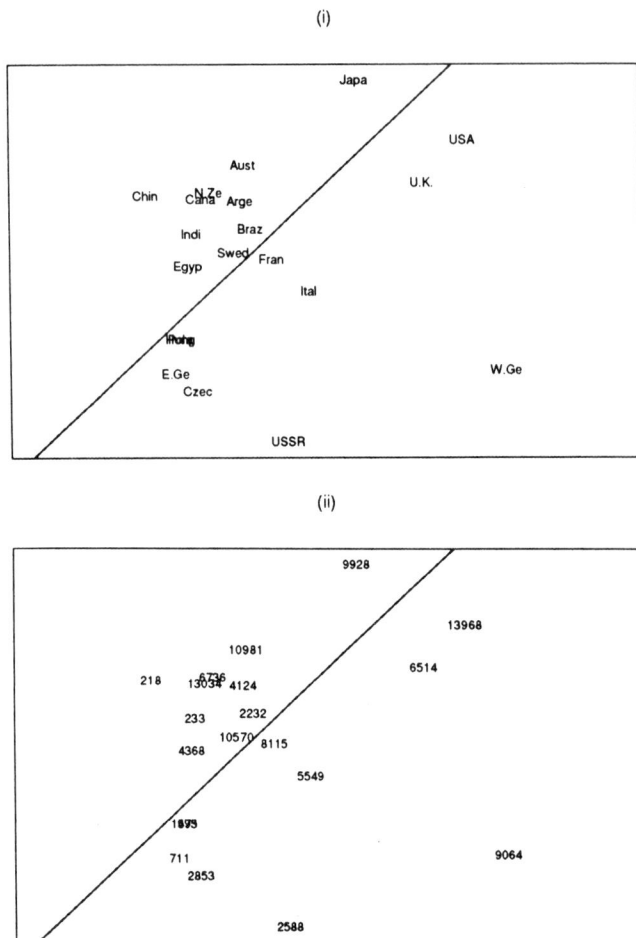

Figure 3.12 *Nonmetric MDS of trading data, together with the gross domestic product per capita axis.*

As long as the regression has a reasonable fit, tested either by an analysis of variance, or by the multiple correlation coefficient, then an axis for the variable can be defined through the origin of the configuration and using the direction cosines

$$\hat{\beta}_i / \sqrt{\sum \hat{\beta}_i^2} \quad (i = 1, \ldots, p).$$

Example

Data were taken from the *New Geographical Digest* (1986) on which countries traded with which other countries. Twenty countries were chosen and their main trading partners were noted. If significant trade occurred between country r and country s, then x_{rs} was put equal to unity, and zero otherwise. From these binary data dissimilarities were calculated using the Jaccard coefficient. Also recorded for the various countries was the gross national product per capita (gnp/cap). The dissimilarities were then subjected to two dimensional nonmetric MDS. Figure 3.12(i) shows the final configuration, the stress for which was 11%, implying a reasonable fit. From the figure, Japan, the USA and the UK can be seen to be separated from the bulk of the countries, while West Germany (as it was then) and USSR (also now changed) are also singled out.

The variable gnp/cap was regressed on the coordinates of the points in the final MDS configuration giving an adjusted coefficient of multiple determination of 42%. Although the fit of gnp/cap is not particularly good, a meaningful axis arises using the regression coefficients to define its direction. This is also shown in Figure 3.12(i). Shown in Figure 3.12(ii) is the same plot but with gnp/cap replacing country names. There are interesting positions taken up by some of the countries, for example Sweden, Canada and the UK. The reader is invited to delve further into the plots.

CHAPTER 4

Further aspects of
multidimensional scaling

4.1 Robust MDS

Spence and Lewandowsky (1989) consider the effect of outliers
in multidimensional scaling. They illustrate the potentially disastrous effects of outliers by using as dissimilarities, the forty-five
Euclidean distances obtained from nine points in a two dimensional
Euclidean space. One of the distances however is multiplied by a
factor of ten. The resulting configuration using classical scaling has
the two points associated with the outlying distance forced well
away from their true positions. To overcome the effects of outliers,
Spence and Lewandowsky suggest a method of robust parameter
estimation and a robust index of fit, described briefly below.

Robust parameter estimation
Suppose a configuration of n points is sought in a p dimensional
Euclidean space with associated distances $\{d_{rs}\}$ representing dissimilarities $\{\delta_{rs}\}$. As usual let the coordinates of the points in the
space be denoted by $\{x_{ri}\}$. Consider the distance between the rth
and sth points,

$$d_{rs}^2 = \sum_{i=1}^{p}(x_{ri} - x_{si})^2.$$

Concentrating on the coordinate x_{rk}, this enters into $n - 1$ distances, and $n - 1$ discrepancies, $f(x_{rk})$, between dissimilarity and
distance,

$$f_s(x_{rk}) = \delta_{rs} - \left\{ \sum_{i=1}^{p}(x_{ri} - x_{si})^2 \right\}^{\frac{1}{2}} \qquad (s \neq r, \ s = 1,\ldots,n).$$

Obviously $f_s(x_{r1}) \equiv f_s(x_{r2}) \equiv \ldots \equiv f_s(x_{rn})$.
    Let $\{x_{ri}^t\}$ be the coordinates at the tth iteration in the search

for the optimum configuration, and let $\{d_{rs}^t\}$ be the associated distances. The Newton-Raphson method for finding roots of equations leads to

$$x_{rk}^{t+1} = x_{rk}^t - \frac{f_s(x_{rk}^t)}{f_s'(x_{rk}^t)} \qquad (s \neq r, \ s = 1, \ldots, n)$$

$$= x_{rk}^t + \frac{(\delta_{rs} - d_{rs}^t)d_{rs}^t}{x_{rk}^t - x_{sk}^t}$$

$$= x_{rk}^t + {}_s g_{rk}^t.$$

The corrections ${}_s g_{rk}^t$ to x_{rk}^t can be greatly influenced by outliers and hence Spence and Lewandowsky suggest using their median. Thus

$$x_{rk}^{t+1} = x_{rk}^t + {}_M g_{rk}^t,$$

where ${}_M g_{rk}^t = \text{median}_{r \neq s}({}_s g_{rk}^t)$.

They also suggest a modification to step size giving

$$x_{rk}^{t+1} = x_{rk}^t + \beta^t \, {}_M g_{rk}^t,$$

where

$$\beta^t = \frac{\alpha^t}{g^t},$$

$$\alpha^{t+1} = \alpha^t \left\{ \frac{\sum_{r,j} (x_{rj}^{t-1} - x_{rj}^{t-2})^2}{\sum_{r,j} (x_{rj}^t - 2x_{rj}^{t-1} + x_{rj}^{t-2})^2} \right\}^{\frac{1}{2}},$$

$$g^t = \left\{ \frac{\sum_{r,j} ({}_M g_{rj}^t)^2}{\sum_{r,j} (x_{rj}^t)^2} \right\}^{\frac{1}{2}}.$$

The above is easily modified to incorporate transformations of distances, dissimilarities or both.

Care has to be taken over the starting configuration. It is suggested that a starting configuration is found by replacing the dissimilarities by their ranks and using classical scaling on these; the configuration is then suitably scaled.

Robust index of fit

Spence and Lewandowsky suggest TUF as an index of fit where

$$\text{TUF} = \text{median}_r \text{median}_{s \neq r} \left| \frac{\delta_{rs} - d_{rs}}{\delta_{rs}} \right|,$$

which when multiplied by 100 can be interpreted as the median percentage discrepancy between the dissimilarities and the fitted distances.

Spence and Lewandowsky carried out simulation exercises to compare several MDS programs in their ability to cope with outliers. They showed that nonmetric methods were more resistant to outliers than metric methods as expected, but that their method (TUFSCAL) was the most resistant.

4.2 Interactive MDS

In an experiment where a subject is presented with pairs of stimuli in order to elicit a dissimilarity/similarity measurement, the number of possible pairs that have to be judged soon becomes overwhelmingly large. An experimental design using just a subset of the possible pairs can be attempted. Spence and Domoney (1974) carried out a Monte Carlo simulation study to show that up to two-thirds of the full set of dissimilarities can be discarded without a disastrous effect on the MDS results.

Young and Cliff (1972) introduce an interactive classical scaling method, where an initial number, n_1, of stimuli are presented for paired comparison. From the resulting scaled dissimilarities the pair of stimuli furthest apart are used to start the definition of a "frame". The frame starts in one dimension as a line passing through two points representing these two stimuli. The distance between the points represents the associated dissimilarity. The rest of the stimuli in the initial set are considered in turn. If the two dissimilarities between a particular stimulus and the two existing frame stimuli can be represented by the distances from the two points to another collinear point, then the stimulus is in the same dimension as the other two. If not a new dimension is required for the frame. The stimulus giving the lowest "residual" distance is used to define another dimension and increases the frame. This process is continued, looking at distances of new stimuli to frame points in terms of projections onto the frame dimensions until a frame of r dimensions is found. Those original stimuli of the n_1 not in the frame are set aside.

More stimuli are added to the ones in the frame and the process is repeated updating the frame. This continues until all the stimuli have been considered and a final frame settled upon. Dissimilarities are then found between those stimuli outside the frame and those within it. Some of these will already have been found however as the frame was being constructed.

Girard and Cliff (1976) carried out a Monte Carlo study to investigate the accuracy of the interactive scaling method by comparing results with a solution based on all the possible dissimilarities, and also with solutions based on subsets of the dissimilarities. They concluded that interactive scaling was superior to simply using a subset of the dissimilarities. Interactive MDS was further developed by Cliff *et al.* (1977) and improved by Green and Bentler (1979).

4.3 Dynamic MDS

Ambrosi and Hansohm (1987) describe a dynamic MDS method for analysing proximity data for a set of objects where dissimilarities are measured at each of T successive time periods. Let these dissimilarities be denoted by $\{\delta_{rs}^t\}$, $(r, s = 1, \ldots, n; t = 1, \ldots, T)$. The aim is to produce a configuration of nT points in a space, where each object is represented T times, once for each of the time periods. The T points for each object are hopefully not too distant from one another and by plotting their path over time, insight into the changing nature of the relationship among the objects with respect to time can be found.

One possibility for coping with the T sets of dissimilarities is to place them into a super-dissimilarity matrix, $\mathbf{D}$,

$$\mathbf{D} = \begin{bmatrix} \mathbf{D}_{11} & \mathbf{D}_{12} & \ldots & \mathbf{D}_{1T} \\ \vdots & \vdots & \ddots & \vdots \\ \mathbf{D}_{T1} & \mathbf{D}_{T2} & \ldots & \mathbf{D}_{TT} \end{bmatrix},$$

where $\mathbf{D}_{tt} = [\delta_{rs}^t]$, the dissimilarity matrix formed from the dissimilarities collected at the tth time period. The matrix $\mathbf{D}_{tt'} = [\delta_{rs}^{t,t'}]$ has to be specified where $\delta_{rs}^{t,t'}$ is the dissimilarity of object r at the tth time period with object s at the t'th time period $(t \neq t')$. Some information may be available from which these cross time period dissimilarities can be found. For example if data matrices were available for the objects, with one for each time period, these dissimilarities could be found using the observations on object r at time period t and those on object s at time period t' to define $\delta_{rs}^{t,t'}$ by the Jaccard coefficient for instance. Usually $\delta_{rs}^{t,t'} \neq \delta_{rs}^{t',t}$ $(r \neq s)$. However the super-dissimilarity matrix will still be symmetric. If the dissimilarities $\delta_{rs}^{t,t'}$ cannot be found, it may be that they can be constructed from $\{\delta_{rs}^t\}$. One possibility is

$$\delta_{rs}^{t,t'} = \tfrac{1}{2}(\delta_{rs}^t + \delta_{rs}^{t'}).$$

Another possibility is to assume all $\delta_{rs}^{t,t'}$ $(t \neq t')$ are missing. A third is to define $\delta_{rr}^{t,t'} = 0$, with all $\delta_{rs}^{t,t'}$ $(r \neq s)$ missing.

Once the super-dissimilarity matrix has been constructed it can be subjected to metric or nonmetric multidimensional scaling in the usual manner.

A different approach is suggested by Ambrosi and Hansohm. They use stress for nonmetric MDS based on the dissimilarities for the tth time period defined by

$$S^t = \frac{\sum_{r<s}(\delta_{rs}^t - \hat{d}_{rs}^t)^2}{\sum_{r<s}(\hat{d}_{rs}^t - \bar{d}^t)^2},$$

where

$$\bar{d} = \frac{2}{n(n-1)} \sum_{r<s} \hat{d}_{rs}^t.$$

The combined stress for the T time periods can be chosen as either

$$S = \frac{\sum_{t=1}^T \sum_{r<s}(\delta_{rs}^t - \hat{d}_{rs}^t)^2}{\sum_{t=1}^T \sum_{r<s}(\hat{d}_{rs}^t - \bar{d}^t)^2},$$

or

$$S = \sum_{t=1}^T S^t.$$

This overall stress is to be minimized, but subject to the constraint that in the resulting configuration, the T points that represent each object tend to be close to each other. This is achieved by using a penalty function, for example

$$U = \sum_{t=1}^{T-1} \sum_{r=1}^n \sum_{i=1}^p (x_{ri}^{t+1} - x_{ri}^t)^2,$$

where $\mathbf{x}_r^t = (x_{ri}^t, \ldots, x_{rp}^t)$ are the coordinates representing object r at the tth time period.

A configuration is then found that minimizes

$$S_\epsilon = S + \epsilon U, \qquad \epsilon > 0,$$

where ϵ is a chosen constant $\ll 1$.

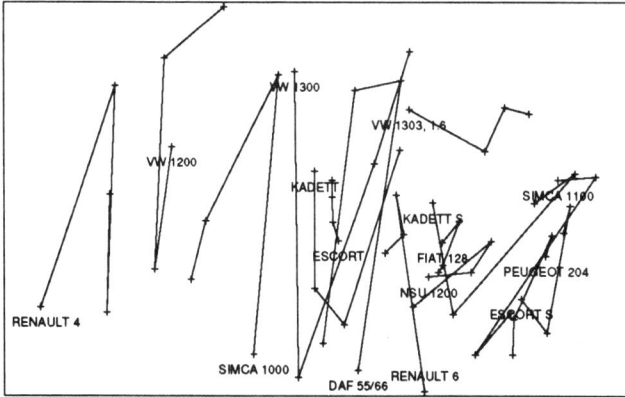

Figure 4.1 *Dynamic MDS for cars using DMDS.*

Minimizing the stress S and also minimizing the penalty function U is then a compromise which will depend on the value of ϵ, which in turn will depend on the importance placed on the requirement that the T points representing an object are near to each other.

A further restriction can be added that the T points representing each object lie on a straight line. This is achieved by insisting

$$\mathbf{x}_r^t = \mathbf{x}_r^1 + \boldsymbol{\alpha}_r^T y_r^t \qquad (r = 1, \ldots, n; t = 2, \ldots, T),$$

where $\boldsymbol{\alpha}_r$ (note in the equation above the superscript T is transpose) gives the direction of the line for the rth object, $\mathbf{x}_r^1$ the starting point of the line, and y_r^t the distance along of the point $\mathbf{x}_r^t$. These new parameters are estimated in the course of minimizing S_ϵ.

An example

Hansohm (1987) describes a computer package DMDS which carries out dynamic MDS. It also includes programs for ordinary MDS, and FACALS, a principal component analysis program. Hansohm illustrates DMDS using data collected by Schobert (1979) on fifteen cars, where each is described by fifteen variables. Data are

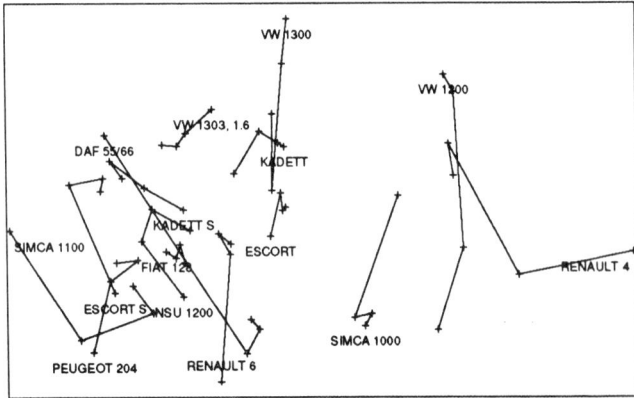

Figure 4.2 *Dynamic MDS for cars using Procrustes analysis.*

collected yearly for the period 1970-1973. Hansohm gives a two dimensional configuration showing the fifteen cars at the four time
points with the points for each car lying on a straight line. The
same data have been used without this restriction and the resulting two dimensional configuration is shown in Figure 4.1. The value
of ϵ was chosen as 0.001, the final penalized stress value was 11%.
The VW cars are on the right of the configuration. The Renault 4
is at the extreme left of the configuration. This, the VW 1200, the
Simca 1000 and the VW 1300 change their positions dramatically
over the time points. Cars to the right of the configuration change
positions less dramatically.

A different approach can be taken to dynamic MDS, by simply
carrying out an MDS analysis for each time period separately, and
then matching the resulting configurations using a Procrustes analysis. This was done for the car data resulting in the configuration
given in Figure 4.2. The stresses for the four initial configurations
were 12%, 7%, 11% and 13%. The configurations for the second,
third and fourth time periods were matched to that of the first.
The values of the Procrustes statistic were 0.29, 0.51 and 0.51 respectively. Alternatively the second configuration could have been
matched to the first, the third to the second, and the fourth to the

third. The length of the trajectories are shorter for this method but cannot be controlled as they can for the previous method by choice of ϵ.

4.4 Constrained MDS

Sometimes it is desirable to place restrictions on the configuration obtained from an MDS analysis, either through parameters or on the distances in the resulting configurations. For example a particular set of stimuli may fall into ten subsets and it is required that all the projections of stimuli points in a subset onto a particular axis are coincident. Bentler and Weeks (1978) describe a situation involving nine Munsell colours of the same red hue, but of differing brightness and saturation, the data coming from Torgerson (1958). The MDS configuration can be constrained so that the first two axes give the true brightness and saturation values for the nine colours.

Another colour example is the data of Ekman (1954) consisting of similarities for fourteen colours. A two dimensional MDS analysis of the data give the colours lying close to the circumference of a circle – the colour circle. Constrained MDS methods can ensure that the colours actually lie on the circumference.

In order to constrain an MDS configuration, Bentler and Weeks (1978) use least squares scaling with the configuration in a Euclidean space and simply incorporate the required equality constraints in the least squares loss function. Bloxom (1978) has the same approach but allows for non-orthogonal axes. Lee and Bentler (1980) also constrain configurations using least squares scaling incorporating Lagrange multipliers. Lee (1984) uses least squares scaling to allow not only for equality constraints but also inequality constraints. Borg and Lingoes (1980) constrain configurations using the following approach which covers the metric and nonmetric methods.

Let $\{\delta_{rs}\}$, $\{d_{rs}\}$ be the usual dissimilarities and distances within a configuration. Let $\{\delta_{rs}^R\}$ be pseudo-dissimilarities which reflect the constraints required. Many of the pseudo-dissimilarities may be missing if they are not involved with constraints. Let $\{\hat{d}_{rs}\}$ be disparities for $\{\delta_{rs}\}$ and $\{\hat{\delta}_{rs}^R\}$ disparities for $\{\delta_{rs}^R\}$ where "disparities" can be the disparities from nonmetric MDS or actual dissimilarities for metric MDS. This allows both cases to be covered simultaneously. Then the constrained solution is found by minimizing the loss function

$$L = (1 - \alpha)L_U + \alpha L_R \qquad (0 \le \alpha \le 1),$$

with

$$L_U = \sum_{r,s}(d_{rs} - \hat{\delta}_{rs})^2,$$

$$L_R = \sum_{r,s}(d_{rs} - \hat{\delta}_{rs}^R)^2.$$

The loss functions L_U and L_R can be Kruskal's STRESS or just a least squares loss function. The loss function L is minimized iteratively with α^t the value of α at the tth iteration. By ensuring that $\lim_{t\to\infty}\alpha^t = 1$, a configuration with the required restrictions is found. Note minimizing L is not the same as minimizing L_R since L_R will contain many missing values while L_U will be complete. Like many other authors Borg and Lingoes use their method on Ekman's colour data and constrain the colours to lie on a circle.

Ter Braak (1992) considers constraining MDS models with regression models, so that coordinates of the configuration are regressed on external variables. He gives as an example a PCO analysis of twenty-one colonies of butterfly where coordinates are regressed on eleven environmental variables. One further constrained MDS model, CANDELINC, will be covered in Chapter 11.

4.4.1 Spherical MDS

Cox and Cox (1991) show how points of a configuration from nonmetric MDS can be forced to lie on the surface of a sphere. In a sense this is not constrained MDS since the space representing the objects is simply taken to be the two dimensional surface of a sphere. The advantage of using the surface of a sphere as a space in which to represent the objects is that the configuration need not have any "edge points", whereas in a Euclidean space there always have to be points at the edge of the configuration. These could be defined as those points lying in the convex hull of the configuration for instance.

The metric methods of constrained MDS of Bloxom (1978), Bentler and Weeks (1978), Lee and Bentler (1980) and Lee (1984) can produce configurations of points lying on the surface of a sphere as particular cases. The nonmetric method of Borg and Lingoes

(1980) can also produce points on a sphere, but is much more awkward than starting with the sphere's surface as space within which to work as with Cox and Cox.

Let the coordinates of the points in the spherical configuration be given by

$$(1, \theta_{1r}, \theta_{2r}) \qquad (r = 1, \ldots, n).$$

Transforming to Cartesian coordinates these are

$$(\cos\theta_{1r}\sin\theta_{2r}, \sin\theta_{1r}\sin\theta_{2r}, \cos\theta_{1r}).$$

The distance between points r and s, d_{rs}, is defined as the shortest arc length along the great circle which passes through the two points. This arc length is monotonically related to the Euclidean distance between the two points (i.e. passing through the interior of the sphere). Since only the rank order of the dissimilarities is important, and hence the rank order of the distances, using the more convenient Euclidean distance rather than the arc length makes very little difference to the resulting configuration. The Euclidean distance is

$$d_{rs} = \{2 - 2\sin\theta_{2r}\sin\theta_{2s}\cos(\theta_{1r} - \theta_{1s}) - 2\cos\theta_{2r}\cos\theta_{2s}\}^{\frac{1}{2}}.$$

Kruskal's stress is defined in the usual manner and then minimized with respect to $\{\theta_{1r}\}$ and $\{\theta_{2r}\}$. The gradient term can be found in Cox and Cox (1991).

The resulting configuration is not unique since an arbitrary rotation of the points or negating one of θ_1 or θ_2 will preserve distances on the sphere and hence give another solution with minimum stress. In passing, note that d_{rs} is invariant to the addition of an arbitrary angle α to each θ_{1r}, but not to the addition of an arbitrary angle β to each θ_{2r}. To find the mapping for an arbitrary rotation first rotate about the z-axis and then the y-axis. This gives

$$(\cos\theta_{1r}\sin\theta_{2r}, \sin\theta_{1r}\sin\theta_{2r}, \cos\theta_{2r}) \rightarrow$$
$$(\cos(\theta_{1r} + \alpha)\sin\theta_{2r}\cos\beta - \cos\theta_{2r}\sin\beta, \sin(\theta_{1r} + \alpha)\sin\theta_{2r},$$
$$\cos(\theta_{1r} + \alpha)\sin\theta_{2r}\sin\beta + \cos\theta_{2r}\cos\beta).$$

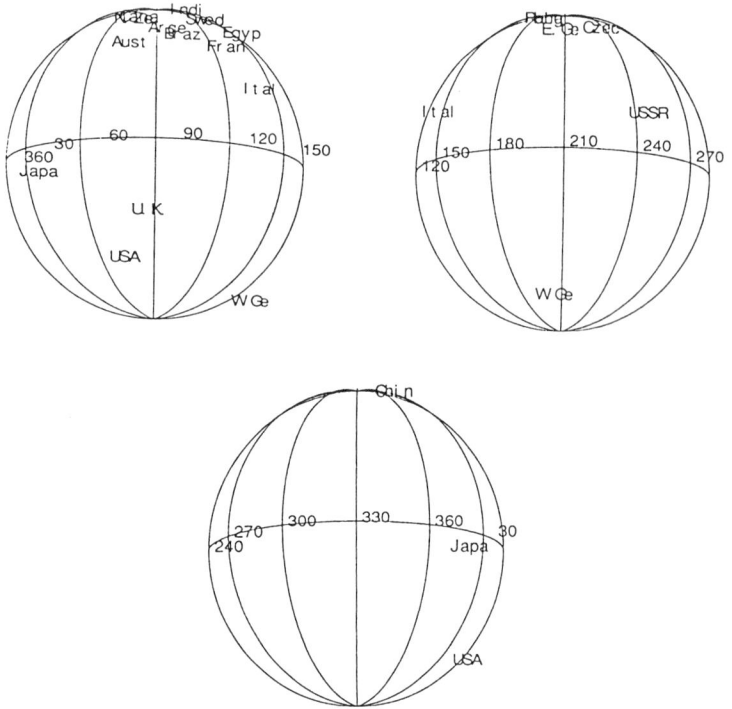

Figure 4.3 *Spherical MDS of the nation trading data.*

An example

The trading data described in the previous chapter were subjected to spherical MDS. The stress for the configuration was 7%, which is 3% less than that for conventional MDS of the data. Figure 4.3 shows the results of subjecting the dissimilarities to spherical MDS. Three views of the sphere are given. Various clusters of countries can be seen, noting of course that there have been political changes since the data were collected. The clusters are {Czechoslovakia, East Germany, Hungary, Poland}, {China, Italy}, {Japan, USA, UK}, {Argentina, Australia, Brazil, Canada, Egypt, France, India, New Zealand, Sweden}, {West Germany}, and {USSR}.

4.5 Statistical inference for MDS

Ramsay (1982) read a paper to the Royal Statistical Society entitled "Some Statistical Approaches to Multidimensional Scaling Data". The content of the paper was the culmination of research into the modelling of dissimilarities incorporating an error structure which leads onto inferential procedures for multidimensional scaling; see Ramsay (1977, 1978a, 1978b, 1980). There followed an interesting discussion with protagonists for and against the use of inference in multidimensional scaling. For instance C. Chatfield said

...and I suggest that this is one area of Statistics [MDS] where the emphasis should remain with data-analytic, exploratory techniques.

B.W. Silverman said

I must say that I am in agreement with Dr. Chatfield in being a little uneasy about the use of multidimensional scaling as a model-based inferential technique, rather than just an exploratory or presentational method.

On the other hand E.E. Roskam said

For a long time, there has been a serious need for some error theory,...

D.R. Cox said

Efforts to discuss some probabilistic aspects of methods that are primarily descriptive are to be welcomed,...

Since 1982 some inferential research has been applied to multidimensional scaling but to date has not made a large impact on the subject. A brief description of some of the inferential ideas is given.

Suppose there is an underlying configuration of points in a Euclidean space that represent objects. As usual let the Euclidean distances between pairs of points be $\{d_{rs}\}$. Let the observed dissimilarity between objects r and s, conditioned on d_{rs}, have probability density function $p(\delta_{rs}|d_{rs})$. It is assumed that these conditioned observations are independent and identically distributed and hence the log-likelihood is

$$l = \sum_r \sum_s \ln p(\delta_{rs}|d_{rs}).$$

The distances can be written in terms of the coordinates of the points, $d_{rs}^2 = (\mathbf{x}_r - \mathbf{x}_s)^T(\mathbf{x}_r - \mathbf{x}_s)$, and hence the log-likelihood can be minimized with respect to $\mathbf{x}_r$ and any parameters of the probability density function p. This gives the maximum likelihood estimates of the coordinates, $\hat{\mathbf{x}}_r$.

Two possible distributions for $\delta_{rs}|d_{rs}$ are the normal and lognormal. For the normal distribution

$$\delta_{rs} \sim \mathrm{N}(d_{rs}, d_{rs}^2 \sigma^2),$$

having constant coefficient of variation. There is a non-zero probability of negative δ_{rs} with this model. For the log-normal distribution

$$\ln \delta_{rs} \sim \mathrm{N}(\ln d_{rs}, \sigma^2).$$

It is possible that a transformation of the dissimilarities is desirable before applying the error structure, such as a power law. Ramsay (1982) suggests a transformation based on monotone splines. However the overall resulting model is rather complicated and this was one of the main reasons it attracted criticism from the discussants of Ramsay's paper.

Once a likelihood has been formulated for multidimensional scaling, further inferences can ensue such as the testing of hypotheses and the formation of confidence regions. For example, if l_k is the maximum value of the log-likelihood when a k dimensional space is used for the configuration of points, then the quantity

$$2(l_k - l_{k-1}),$$

has an asymptotic χ^2 distribution with $n - k$ degrees of freedom, where n is the number of points in the configuration. This can be used to assess the required number of dimensions needed.

Takane (1978a,b) introduces a maximum likelihood method for nonmetric scaling. Let there be an underlying configuration of points representing the objects with coordinates $\{\mathbf{x}_r\}$ and distances between points $\{d_{rs}\}$. For an additive error model let there be a latent variable λ_{rs} so that

$$\lambda_{rs} = d_{rs} + \epsilon_{rs}, \qquad \epsilon_{rs} \sim \mathrm{N}(0, \sigma_{rs}^2).$$

Then if $\lambda_{rs} \geq \lambda_{r's'}$, for the observed dissimilarities $\delta_{rs} \succ \delta_{r's'}$ where $\succ$ represents ordering of the dissimilarities, define

$$\begin{aligned} Y_{rsr's'} &= 1 && \text{if } \delta_{rs} \succ \delta_{r's'} \\ &= 0 && \text{if } \delta_{rs} \prec \delta_{r's'}. \end{aligned}$$

Then

$$\Pr(Y_{rsr's'} = 1) = \Pr(\lambda_{rs} - \lambda_{r's'} \geq 0)$$

$$= \Phi\left(\frac{d_{rs} - d_{r's'}}{(\sigma_{rs}^2 + \sigma_{r's'}^2)^{\frac{1}{2}}}\right) \qquad (= \Phi_{rsr's'} \text{ say})$$

and hence the likelihood is given by

$$L = \prod \Phi_{rsr's'}^{Y_{rsr's'}} (1 - \Phi_{rsr's'})^{1-Y_{rsr's'}},$$

assuming independence of $\{\epsilon_{rs}\}$.

Writing d_{rs} in terms of the coordinates $\mathbf{x}_r$ allows the likelihood or log-likelihood to be maximized with respect to these, giving $\hat{\mathbf{x}}_r$ as the maximum likelihood configuration.

In a similar manner Takane (1981) gives a maximum likelihood approach to multidimensional successive categories scaling. Successive categories scaling is a special case where dissimilarities are ordered categorical variables. So for the whisky tasting experiment of Chapter 1, the possible categories for comparison of two whiskies might be: very similar; similar; neutral (neither similar nor dissimilar); dissimilar; and very dissimilar. The categories could be assigned scores 0, 1, 2, 3, 4 and hence the dissimilarities $\{\delta_{rs}\}$ can each take one of only five possible values. The dissimilarities could then be subjected to metric or nonmetric MDS in the usual manner. Takane suggests the following model, assuming a single set of dissimilarities. Again let there be an underlying configuration of points representing the objects, with coordinates $\mathbf{x}_r$ and distances between points d_{rs}, an additive error. Let there be a latent variable λ_{rs} as above. The successive categories are represented by a set of ordered intervals

$$-\infty = b_0 \leq b_1 \leq \ldots \leq b_M = \infty,$$

where the number of possible categories is M. So the interval $(b_{i-1}, b_i]$ represents the ith category. If the value of the latent variable λ_{rs} lies in the interval $(b_{i-1}, b_i]$ then δ_{rs} is observed as being in the ith category.

Let p_{rsi} be the probability that δ_{rs} is observed as being in the ith category. Then

$$p_{rsi} = \Phi\left(\frac{b_i - d_{rs}}{\sigma}\right) - \Phi\left(\frac{b_{i-1} - d_{rs}}{\sigma}\right).$$

Let the indicator variable Z_{rsi} be defined as

$$Z_{rsi} = 1 \quad \text{if } \delta_{rs} \text{ falls in the } i\text{th category}$$
$$0 \quad \text{otherwise}$$

Assuming independence, the likelihood of $\{Z_{rsi}\}$ is then given by

$$L = \prod_r \prod_s \prod_i p_{rsi}^{z_{rsi}}$$

and hence the log-likelihood is

$$l = \sum_r \sum_s \sum_i z_{rsi} \ln p_{rsi}.$$

The log-likelihood is then maximized with respect to the category boundaries $\{b_i\}$, the coordinates $\{\mathbf{x}_r\}$ and the error variance σ^2. This then gives the maximum likelihood configuration $\{\hat{\mathbf{x}}_r\}$. The procedure can easily be generalized to the cases of replications and several judges.

Zinnes and MacKay (1983) report on a different approach for introducing probabilistic errors, using the Hefner model (Hefner, 1958). Here each stimulus (conceptually it is easier to think of stimuli rather than objects) is represented by a p dimensional random vector $\mathbf{X}_r = (X_{r1}, \ldots, X_{rp})^T$. All components, X_{ri}, of $\mathbf{X}_r$ are assumed independently normally distributed with mean μ_r and variance σ_r^2. These distributions then induce a distribution on the Euclidean distance $(\mathbf{X}_r - \mathbf{X}_s)^T(\mathbf{X}_r - \mathbf{X}_s)$, and it is assumed that the observed dissimilarity is this Euclidean distance. Thus

$$\delta_{rs} = \{(\mathbf{X}_r - \mathbf{X}_s)^T(\mathbf{X}_r - \mathbf{X}_s)\}^{\frac{1}{2}}.$$

It is also assumed that the "true" distance between points r and s is given by

$$d_{rs}^2 = (\boldsymbol{\mu}_r - \boldsymbol{\mu}_s)^T(\boldsymbol{\mu}_r - \boldsymbol{\mu}_s),$$

where $\boldsymbol{\mu} = (\mu_{r1}, \ldots, \mu_{rp})^T$, $(r = 1, \ldots, n)$.

Hefner (1958) has shown that $\delta_{rs}^2/(\sigma_r^2 + \sigma_s^2)$ has a non-central chi-squared distribution, $\chi'^2(p, d_{rs}^2/(\sigma_r^2 + \sigma_s^2))$. From this it is possible to find the distribution of δ_{rs}. Zinnes and MacKay give approximations to the probability density functions. For $(d_{rs}\delta_{rs})/(\sigma_r^2 + \sigma_s^2) \geq 2.55$ the density function can be approximated by

$$2\delta_{rs}^{-1}z\phi\left[\left(\frac{z}{p+\lambda}\right)^h\right](p+\lambda)^{-h}(hz^{h-1}),$$

where $z = \delta_{rs}^2/(\sigma_r^2 + \sigma_s^2)$, $\lambda = d_{rs}^2/(\sigma_r^2 + \sigma_s^2)$, $h = 1 - \frac{2}{3}(p + \lambda)(p + 3\lambda)/(p + 2\lambda)^2$.

For $(d_{rs}\delta_{rs})/(\sigma_r^2 + \sigma_s^2) < 2.55$ an approximation based on beta functions is used,

$$2\delta_{rs}^{-1} \exp\{-\tfrac{1}{2}(z^2 + \lambda)\}(\tfrac{1}{2}z)^{p/2} \sum_{k=0}^{\infty} A_k,$$

where

$$A_0 = \frac{1}{\Gamma}\frac{p}{2}, \quad A_k = \frac{\lambda z}{4k(k + \frac{1}{2}p - 1)} A_{k-1},$$

with five terms in the summation usually giving sufficient accuracy.

Zinnes and MacKay maximize the sum of the logarithms of the approximating density functions, one for each dissimilarity, with respect to $\{\mu_r\}$ and $\{\sigma_r^2\}$. The values $\{\mu_r\}$ give the coordinates of the points in the configuration.

Brady (1985) considered in detail consistency and hypothesis testing for nonmetric MDS. Brady's work is very general and he uses his own special notation. Bennett (1987) considers influential observations in multidimensional scaling.

4.6 Asymmetric dissimilarities

Metric and nonmetric MDS methods so far described have been for one-mode, two-way symmetric data, where the symmetry in dissimilarities (similarities) $\delta_{rs} = \delta_{sr}$ is reflected in the symmetry in distances within the MDS configuration $d_{rs} = d_{sr}$. Some situations give rise to asymmetric proximities. For example within a school class, each child is asked to score the friendship he/she feels for each of the other members of the class. Results are unlikely to be symmetric.

In the early days of MDS, Kruskal (1964a) suggested two approaches that could be taken with asymmetric dissimilarities. The first was to average δ_{rs} and δ_{sr} and proceed as usual. The other was to let the summations in STRESS extend over all $r \neq s$ rather than $r < s$. Another possibility is to represent every object twice with new dissimilarities $\delta'_{ni+r,nj+s}$ $(i, j = 0, 1)$, where both r and $n + r$ represent the rth object. Let

$$\delta'_{r,s} = \delta'_{s,r} = \delta_{rs}$$
$$\delta'_{n+r,n+s} = \delta'_{n+s,n+r} = \delta_{sr}$$
$$\delta'_{r,n+r} = \delta'_{n+r,r} = 0$$

and treat dissimilarities $\delta'_{r,n+s}$ and $\delta_{n+r,s}$ $(r \neq s)$ as missing.

The above methods attempt to overcome the problem of asymmetric dissimilarities using techniques designed for symmetric dissimilarities. It is more satisfactory to model the asymmetry. Gower (1977) does this in several ways. Let the dissimilarities be placed in matrix $\mathbf{D}$. His first method is to use the singular value decomposition of $\mathbf{D}$,

$$\mathbf{D} = \mathbf{U} \Lambda \mathbf{V}^T,$$

whereupon

$$\mathbf{V} \mathbf{U}^T \mathbf{D} = \mathbf{V} \Lambda \mathbf{V}^T$$
$$\mathbf{D} \mathbf{V} \mathbf{U}^T = \mathbf{U} \Lambda \mathbf{U}^T$$

are both symmetric and the orthogonal matrix $\mathbf{A} = \mathbf{U}^T \mathbf{V}$ can be regarded as a measure of symmetry, since if $\mathbf{D}$ is symmetric, $\mathbf{A} = \mathbf{I}$.

His second method is as follows. Define matrix $\mathbf{A}$ as $[\mathbf{A}]_{rs} = -\frac{1}{2} d_{rs}^2$. Centre its rows and columns as for classical scaling. Express $\mathbf{A}$ as

$$\mathbf{A} = \mathbf{U} \Lambda \mathbf{U}^{-1},$$

where Λ is the diagonal matrix of eigenvalues of $\mathbf{A}$, some of which could be complex. Matrix $\mathbf{U}$ consists of the left eigenvectors of $\mathbf{A}$, and matrix $\mathbf{U}^{-1}$ the right eigenvectors. Use $\mathbf{U} \Lambda^{\frac{1}{2}}$, $(\mathbf{U}^{-1})^T \Lambda^{\frac{1}{2}}$ to plot two configurations of points. If $\mathbf{A}$ was symmetric the configurations would coincide.

The third method works on the rows and columns of $\mathbf{D}$ expressed in terms of the upper and lower triangular matrices of $\mathbf{D}$, $\mathbf{D} = (\mathbf{L} \backslash \mathbf{U})$. The rows and columns are permuted so that

$$\left| \sum [\mathbf{U}]_{rs} - \sum [\mathbf{L}]_{rs} \right|$$

is a maximum. Then make $\mathbf{U}$ and $\mathbf{L}$ symmetric and subject both to an MDS analysis. The results can be regarded as the worst situation for the asymmetry of $\mathbf{D}$.

Gower's fourth method is simply to use multidimensional unfolding or correspondence analysis on $\mathbf{D}$. See Chapters 7 and 8.

Gower's fifth method searches for the best rank 2 matrix that when added to $\mathbf{D}$ makes it the most symmetric, measured by

$$\sum_{r<s}(\delta_{rs} - \delta_{sr})^2.$$

Weeks and Bentler (1982) propose the following model for asymmetry. Decompose the matrix $\mathbf{D}$ into two matrices, one symmetric and one skew-symmetric,

$$\mathbf{D} = \mathbf{A} + \mathbf{B}$$

where

$$[\mathbf{A}]_{rs} = a_{rs} = a_{sr} = (\delta_{rs} + \delta_{sr})/2,$$
$$[\mathbf{B}]_{rs} = b_{rs} = b_{sr} = (\delta_{rs} - \delta_{sr})/2.$$

Then the dissimilarity δ_{rs} is modelled by

$$\delta_{rs} = \alpha d_{rs} + \beta + c_r - c_s + \epsilon_{rs},$$

where d_{rs} is Euclidean distance, α, β are parameters, c_r, c_s represent the skew-symmetric component, and ϵ_{rs} an error component. Constraints are placed to make the model identifiable and then the various parameters, components and coordinates are estimated using least squares.

Another MDS model for asymmetric dissimilarities is DEDICOM which is described in Chapter 11.

CHAPTER 5

Procrustes analysis

5.1 Introduction

It is often necessary to compare one configuration of points in a
Euclidean space with another, where there is a one-to-one mapping
from one set of points to the other. For instance the configuration
of points obtained from an MDS analysis on a set of objects might
need to be compared with a configuration obtained from a different analysis or perhaps with an underlying configuration such as
physical location.

The technique of matching one configuration to another and
producing a measure of the match is called Procrustes analysis.
This particular technique is probably the only statistical method
to be named after a villain. Any traveller on the road from Eleusis
to Athens in ancient Greece was in for a surprise if he accepted
the kind hospitality and a bed for the night from a man named
Damastes, who lived by the roadside. If his guests did not fit the
bed, Damastes would either stretch them on a rack to make them
fit if they were too short, or chop off their extremities if they
were too long. Damastes earned the nickname Procrustes mean-
ing "stretcher". Procrustes eventually experienced the same fate
as that of his guests at the hands of Theseus – all this of course
according to Greek mythology.

Procrustes analysis seeks the isotropic dilation and the rigid
translation, reflection and rotation needed to best match one configuration to the other. Solutions to the problem of finding these
motions have been given by Green (1952), Schönemann (1966),
Schönemann and Carroll (1970). Sibson (1978) gives a short review of Procrustes analysis and sets out the solution.

5.2 Procrustes analysis

Suppose a configuration of n points in a q dimensional Euclidean space, with coordinates given by the $n \times q$ matrix $\mathbf{X}$, needs to be optimally matched to another configuration of n points in a p $(p \geq q)$ dimensional Euclidean space with coordinate matrix $\mathbf{Y}$. It is assumed that the rth point in the first configuration is in a one-to-one correspondence with the rth point in the second configuration. The points in the two configurations could be representing objects, cities, stimuli, etc. Firstly $p - q$ columns of zeros are placed at the end of matrix $\mathbf{X}$ so that both configurations are placed in p dimensional space. The sum of the distances between the points in the $\mathbf{Y}$ space and the corresponding points in the $\mathbf{X}$ space is given by

$$R^2 = \sum_{r=1}^{n} (\mathbf{y}_r - \mathbf{x}_r)^T (\mathbf{y}_r - \mathbf{x}_r)$$

where $\mathbf{X} = [\mathbf{x}_1, \ldots, \mathbf{x}_n]^T$, $\mathbf{Y} = [\mathbf{y}_1, \ldots, \mathbf{y}_n]^T$, and $\mathbf{x}_r$ and $\mathbf{y}_r$ are the coordinate vectors of the rth point in the two spaces.

Let the points in the $\mathbf{X}$ space be dilated, translated, rotated, reflected to new coordinates $\mathbf{x}'_r$, where

$$\mathbf{x}'_r = \rho \mathbf{A}^T \mathbf{x}_r + \mathbf{b}.$$

The matrix $\mathbf{A}$ is orthogonal giving a rigid rotation, vector $\mathbf{b}$ is a rigid translation vector, and ρ is the dilation. The motions are sought that minimize the new sum of distances between points,

$$R^2 = \sum_{r=1}^{n} (\mathbf{y}_r - \rho \mathbf{A}^T \mathbf{x}_r - \mathbf{b})^T (\mathbf{y}_r - \rho \mathbf{A}^T \mathbf{x}_r - \mathbf{b}). \tag{5.1}$$

Optimal translation
Let $\mathbf{x}_0$, $\mathbf{y}_0$ be the centroids of the two configurations,

$$\mathbf{x}_0 = \frac{1}{n} \sum_{r=1}^{n} \mathbf{x}_r, \quad \mathbf{y}_0 = \frac{1}{n} \sum_{r=1}^{n} \mathbf{y}_r.$$

Measuring $\mathbf{x}_r$ and $\mathbf{y}_r$ relative to these centroids in (5.1) gives

$$R^2 = \sum_{r=1}^{n} \left((\mathbf{y}_r - \mathbf{y}_0) - \rho \mathbf{A}^T (\mathbf{x}_r - \mathbf{x}_0) + \mathbf{y}_0 - \rho \mathbf{A}^T \mathbf{x}_0 - \mathbf{b} \right)^T$$

$$\left((\mathbf{y}_r - \mathbf{y}_0) - \rho \mathbf{A}^T (\mathbf{x}_r - \mathbf{x}_0) + \mathbf{y}_0 - \rho \mathbf{A}^T \mathbf{x}_0 - \mathbf{b} \right).$$

On expanding

$$R^2 = \sum_{r=1}^{n} \left((\mathbf{y}_r - \mathbf{y}_0) - \rho \mathbf{A}^T(\mathbf{x}_r - \mathbf{x}_0) \right)^T$$

$$\left((\mathbf{y}_r - \mathbf{y}_0) - \rho \mathbf{A}^T(\mathbf{x}_r - \mathbf{x}_0) \right)$$

$$+ n \left(\mathbf{y}_0 - \rho \mathbf{A}^T \mathbf{x}_0 - \mathbf{b} \right)^T \left(\mathbf{y}_0 - \rho \mathbf{A}^T \mathbf{x}_0 - \mathbf{b} \right). \quad (5.2)$$

Since the last term in (5.2) is non-negative, and $\mathbf{b}$ only occurs in this term, in order that R^2 be a minimum,

$$\mathbf{b} = \mathbf{y}_0 - \rho \mathbf{A}^T \mathbf{x}_0.$$

Hence

$$\mathbf{x}'_r = \rho \mathbf{A}^T(\mathbf{x}_r - \mathbf{x}_0) + \mathbf{y}_0,$$

which implies the centroid in the $\mathbf{X}'$ space is coincident with the centroid in the Y space. The most convenient way of ensuring this is initially to translate the configurations in the $\mathbf{X}$ space and $\mathbf{Y}$ space, so that they both have their centroids at the origin.

Optimal dilation
Now assuming $\mathbf{x}_0 = \mathbf{y}_0 = \mathbf{0}$, then

$$R^2 = \sum_{r=1}^{n} (\mathbf{y}_r - \rho \mathbf{A}^T \mathbf{x}_r)^T (\mathbf{y}_r - \rho \mathbf{A}^T \mathbf{x}_r)$$

$$= \sum_{r=1}^{n} \mathbf{y}_r^T \mathbf{y}_r + \rho^2 \sum_{r=1}^{n} \mathbf{x}_r^T \mathbf{x}_r - 2\rho \sum_{r=1}^{n} \mathbf{x}_r^T \mathbf{A} \mathbf{y}_r$$

$$= \mathrm{tr}(\mathbf{YY}^T) + \rho^2 \mathrm{tr}(\mathbf{XX}^T) - 2\rho \mathrm{tr}(\mathbf{XAY}^T). \quad (5.3)$$

Differentiating with respect to ρ gives $\hat{\rho}$, the value of ρ giving R^2 as a minimum,

$$\hat{\rho} = \mathrm{tr}(\mathbf{XAY}^T)/\mathrm{tr}(\mathbf{XX}^T),$$

$$= \mathrm{tr}(\mathbf{AY}^T\mathbf{X})/\mathrm{tr}(\mathbf{XX}^T).$$

The rotation matrix, $\mathbf{A}$, is still unknown and needs to be considered next.

Optimal rotation
Sibson (1978) derives the optimal rotation matrix with an elegant

proof not requiring matrix differentiation of R^2. The following is based on his work. For the alternative approach, using matrix differentiation, see for example Mardia *et al.* (1979).

The value of R^2 in (5.3) will be a minimum if $\mathrm{tr}(\mathbf{XAY}^T) = \mathrm{tr}(\mathbf{AY}^T\mathbf{X})$ is a maximum. Let $\mathbf{C} = \mathbf{Y}^T\mathbf{X}$, and let $\mathbf{C}$ have the singular value decomposition

$$\mathbf{C} = \mathbf{U}\boldsymbol{\Lambda}\mathbf{V}^T,$$

where $\mathbf{U}$ and $\mathbf{V}$ are orthonormal matrices and $\boldsymbol{\Lambda}$ is a diagonal matrix of singular values. Then

$$\mathrm{tr}(\mathbf{AC}) = \mathrm{tr}(\mathbf{AU}\boldsymbol{\Lambda}\mathbf{V}^T) = \mathrm{tr}(\mathbf{V}^T\mathbf{AU}\boldsymbol{\Lambda}).$$

Now $\mathbf{V}$, $\mathbf{A}$ and $\mathbf{U}$ are all orthonormal matrices; and hence so is $\mathbf{V}^T\mathbf{AU}$. Since $\boldsymbol{\Lambda}$ is diagonal and an orthogonal matrix cannot have any element greater than unity,

$$\mathrm{tr}(\mathbf{AC}) = \mathrm{tr}(\mathbf{V}^T\mathbf{AU}\boldsymbol{\Lambda}) \leq \mathrm{tr}(\boldsymbol{\Lambda}).$$

Thus R^2 is minimized when $\mathrm{tr}(\mathbf{AC}) = \mathrm{tr}(\boldsymbol{\Lambda})$, implying

$$\mathbf{V}^T\mathbf{AU}\boldsymbol{\Lambda} = \boldsymbol{\Lambda}. \tag{5.4}$$

Equation (5.4) has solution $\mathbf{A} = \mathbf{V}\mathbf{U}^T$, giving the optimal rotation matrix as the product of the orthonormal matrices in the SVD of $\mathbf{Y}^T\mathbf{X}$.

The solution can be taken further. Pre-multiplying and post-multiplying (5.4) by $\mathbf{V}$ and $\mathbf{V}^T$ respectively,

$$\mathbf{AU}\boldsymbol{\Lambda}\mathbf{V}^T = \mathbf{V}\boldsymbol{\Lambda}\mathbf{V}^T.$$

Hence

$$\mathbf{AC} = \mathbf{V}\boldsymbol{\Lambda}\mathbf{V}^T = (\mathbf{V}\boldsymbol{\Lambda}^2\mathbf{V}^T)^{\frac{1}{2}} = (\mathbf{V}\boldsymbol{\Lambda}\mathbf{U}\mathbf{U}^T\boldsymbol{\Lambda}\mathbf{V}^T)^{\frac{1}{2}}$$

$$= (\mathbf{C}^T\mathbf{C})^{\frac{1}{2}}.$$

Thus the optimal rotation matrix is given by

$$(\mathbf{C}^T\mathbf{C})^{\frac{1}{2}}\mathbf{C}^{-1} = (\mathbf{X}^T\mathbf{Y}\mathbf{Y}^T\mathbf{X})^{\frac{1}{2}}(\mathbf{Y}^T\mathbf{X})^{-1},$$

if $\mathbf{Y}^T\mathbf{X}$ is nonsingular and by a solution of

$$\mathbf{AC} = (\mathbf{C}^T\mathbf{C})^{\frac{1}{2}}$$

otherwise. Note that the solution no longer requires the SVD of $\mathbf{Y}^T\mathbf{X}$, which was only needed in the proof of (5.4).

Returning to the optimal dilation, it is now seen that

$$\hat{\rho} = \text{tr}(\mathbf{X}^T \mathbf{Y} \mathbf{Y}^T \mathbf{X})^{\frac{1}{2}} / \text{tr}(\mathbf{X}^T \mathbf{X}).$$

Assessing the match of the two configurations can be done using the minimized value of R^2, which is

$$R^2 = \text{tr}(\mathbf{Y} \mathbf{Y}^T) - \{\text{tr}(\mathbf{X}^T \mathbf{Y} \mathbf{Y}^T \mathbf{X})^{\frac{1}{2}}\}^2 / \text{tr}(\mathbf{X}^T \mathbf{X}).$$

The value of R^2 can now be scaled, for example by dividing by $\text{tr}(\mathbf{Y}^T \mathbf{Y})$ to give

$$R^2 = 1 - \{\text{tr}(\mathbf{X}^T \mathbf{Y} \mathbf{Y}^T \mathbf{X})^{\frac{1}{2}}\}^2 / \{\text{tr}(\mathbf{X}^T \mathbf{X}) \text{tr}(\mathbf{Y}^T \mathbf{Y})\}.$$

This is known as the Procrustes statistic.

5.2.1 Procrustes analysis in practice

Summarizing the steps in a Procrustes analysis where configuration Y is to be matched to configuration X:

1. Subtract the mean vectors for the configurations from each of the respective points in order to have the centroids at the origin.
2. Find the rotation matrix $\mathbf{A} = (\mathbf{X}^T \mathbf{Y} \mathbf{Y}^T \mathbf{X})^{\frac{1}{2}} (\mathbf{Y}^T \mathbf{X})^{-1}$ and rotate the X configuration to $\mathbf{X}\mathbf{A}$.
3. Scale the X configuration by multiplying each coordinate by ρ, where $\rho = \text{tr}(\mathbf{X}^T \mathbf{Y} \mathbf{Y}^T \mathbf{X})^{\frac{1}{2}} / \text{tr}(\mathbf{X}^T \mathbf{X})$.
4. Calculate the minimized and scaled value of

$$R^2 = 1 - \{\text{tr}(\mathbf{X}^T \mathbf{Y} \mathbf{Y}^T \mathbf{X})^{\frac{1}{2}}\}^2 / \{\text{tr}(\mathbf{X}^T \mathbf{X}) \text{tr}(\mathbf{Y}^T \mathbf{Y})\}.$$

5.3 Historic maps

The construction of maps centuries ago was clearly not an easy task, where only crude measuring instruments could be used, in contrast to the satellite positioning available today. John Speed's County Atlas, the *Theatre of the Empire of Great Britain*, was first engraved and printed in Amsterdam by Jodous Hondius in 1611-1612. A copy of his map of Worcestershire appears in Bricker *et al.* (1976). Twenty towns and villages were chosen from the map and their coordinates were found for each by measuring from the lower left hand corner of the map. The corresponding places were also found on the Landranger Series of Ordnance Survey Maps

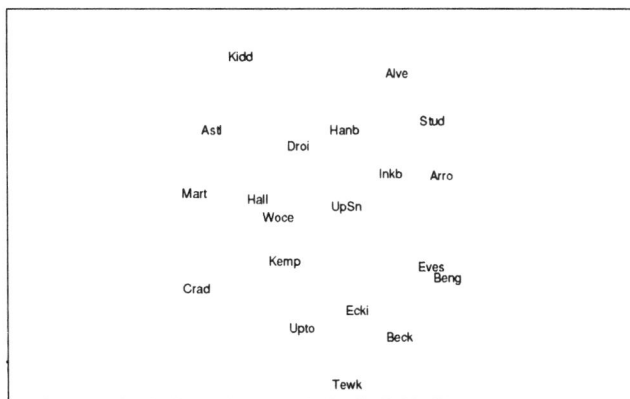

Figure 5.1 *The location of the following villages and towns from Speed's map: Alvechurch, Arrow, Astley, Beckford, Bengeworth, Cradley, Droitwich, Eckington, Evesham, Hallow, Hanbury, Inkberrow, Kempsey, Kidderminster, Martley, Studley, Tewkesbury, Upton, Snodsbury, Worcester.*

(numbers 150,139,138) and their coordinates noted. Since the area covered was relatively small, any projection from the earth's sphere onto a two dimensional plane was ignored. Historic buildings like churches were taken as the point locations of the towns and villages. A Procrustes analysis for the two configurations of places should give some insight into the accuracy of the early map.

Figure 5.1 shows the locations of the various villages and towns from Speed's map. Figure 5.2(i) shows the same places according to Ordnance Survey maps. The configuration of points in Speed's map was subjected to Procrustes analysis giving rise to the rotated, dilated and translated set of points in Figure 5.2(ii). The points did not have to move very far as indicated by the value of the Procrustes statistic, 0.004, indicating that Speed's map was fairly accurate. The root mean squared distance between corresponding points was equivalent to a distance of about 9 miles; a possible measure of the accuracy of the early map.

(i)

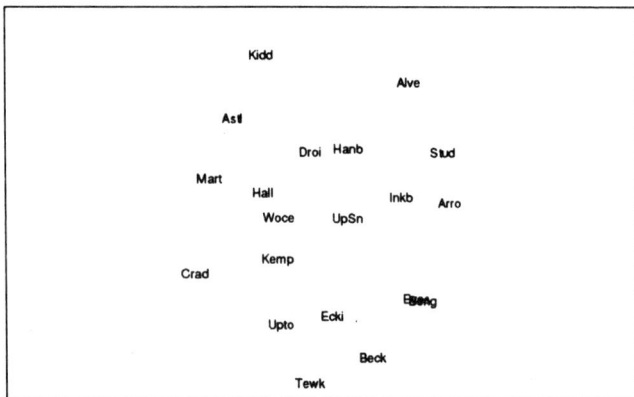

(ii)

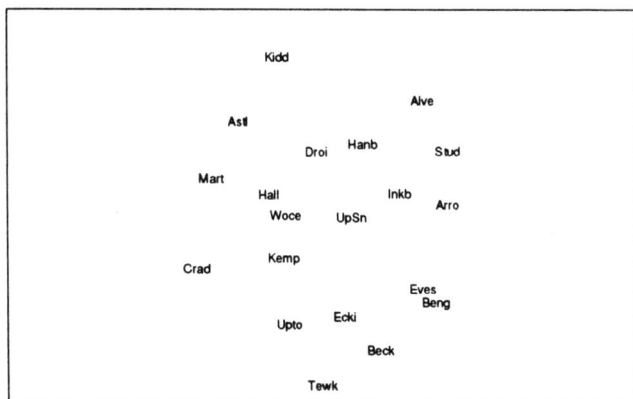

Figure 5.2 *(i) The locations of the villages and towns from Ordnance Survey maps; (ii) Speed's map after Procrustes analysis to match it to the Ordnance Survey map.*

5.4 Some generalizations

Once the configurations of points have been translated to have centroids at the origin, the "Procrustes analysis" or the "Procrustes rotation" described in the previous section can be simply described as the rotation of a matrix $\mathbf{X}$ so that it matches matrix $\mathbf{Y}$ as best as possible. This was achieved by essentially minimizing

$$R^2 = \mathrm{tr}(\mathbf{Y} - \mathbf{XA})^T(\mathbf{Y} - \mathbf{XA}).$$

The technique can be described as the unweighted orthogonal Procrustes rotation. Some generalizations of the method are briefly explained.

5.4.1 Weighted Procrustes rotation

Suppose the contribution to R^2 by point r is to be weighted by an amount ω_r^2 $(r = 1, \ldots, n)$. Then the rotation $\mathbf{A}$ is sought that minimizes

$$R^2 = \mathrm{tr}(\mathbf{Y} - \mathbf{XA})^T\mathbf{W}_n^2(\mathbf{Y} - \mathbf{XA}), \qquad (5.5)$$

where $\mathbf{W}_n = \mathrm{diag}(\omega_1, \ldots, \omega_n)$.

Now R^2 can be rewritten as

$$R^2 = \mathrm{tr}(\mathbf{W}_n\mathbf{Y} - \mathbf{W}_n\mathbf{XA})^T(\mathbf{W}_n\mathbf{Y} - \mathbf{W}_n\mathbf{XA}),$$

and hence the solution for the unweighted case can be used, but using $\mathbf{W}_n\mathbf{X}$ and $\mathbf{W}_n\mathbf{Y}$ instead of $\mathbf{X}$ and $\mathbf{Y}$. Lissitz et al. (1976) is an early reference to the problem. See also Gower (1984).

If instead of weighting the points in the configuration, the dimensions are weighted in matrix $\mathbf{Y}$, then the appropriate quantity to minimize is now

$$R^2 = \mathrm{tr}(\mathbf{Y} - \mathbf{XA})\mathbf{W}_n^2(\mathbf{Y} - \mathbf{XA})^T,$$

where $\mathbf{W}_p = \mathrm{diag}(\omega_1, \ldots, \omega_p)$. This case is much more difficult to solve since $\mathbf{X}$ and $\mathbf{Y}$ cannot simply be replaced by $\mathbf{XW}_p$ and $\mathbf{YW}_p$. Lissitz et al. (1976) show that R^2 can be minimized if the condition $\mathbf{AA}^T = \mathbf{I}$, is replaced by $\mathbf{AW}_p^2\mathbf{A}^T = \mathbf{I}$, but as noted by Gower (1984) and Koschat and Swayne (1991), this may be convenient mathematically but does not solve the original problem.

Mooijaart and Commandeur (1990) and Koschat and Swayne (1991) give a solution to the problem. Following the latter, suppose the column vectors in $\mathbf{X}$ are pairwise orthogonal and each of length ρ, so that $\mathbf{X}^T\mathbf{X} = \rho\mathbf{I}$. Then

$$\begin{aligned}
R^2 &= \mathrm{tr}(\mathbf{Y} - \mathbf{XA})\mathbf{W}_p^2(\mathbf{Y} - \mathbf{XA})^T \\
&= \mathrm{tr}(\mathbf{YW}_p^2\mathbf{Y}^T) - 2\mathrm{tr}(\mathbf{W}_p^2\mathbf{Y}^T\mathbf{XA}) + \rho^2\mathrm{tr}(\mathbf{W}_p^2),
\end{aligned}$$

and hence minimization of R^2 is equivalent to the maximization of

$$\text{tr}(\mathbf{W}_p^2 \mathbf{Y}^T \mathbf{X} \mathbf{A}) = \text{tr}(\mathbf{X} \mathbf{A} \mathbf{W}_p^2 \mathbf{Y}^T)$$

and can be achieved in the same manner as for the unweighted case with Y^T replaced by $\mathbf{W}_p^2 \mathbf{Y}^T$. This result can now be used for general $\mathbf{X}$ as follows.

Enlarge $\mathbf{X}$ and $\mathbf{Y}$ to $(n + p) \times p$ matrices

$$\mathbf{X}^* = \begin{bmatrix} \mathbf{X} \\ \mathbf{X}_a \end{bmatrix}, \qquad \mathbf{Y}_1^* = \begin{bmatrix} \mathbf{Y} \\ \mathbf{Y}_1 \end{bmatrix}$$

with $\mathbf{X}_a$ chosen so that $\mathbf{X}^{*T} \mathbf{X}^* = \rho^2 \mathbf{I}$ for some ρ. Thus $\mathbf{X}_a$ is chosen so that

$$\mathbf{X}_a^T \mathbf{X}_a = \rho^2 \mathbf{I} - \mathbf{X}^T \mathbf{X}.$$

Koschat and Swayne suggest using $\rho = 1.1$ times the largest eigenvalue of $\mathbf{X}^T \mathbf{X}$, and chosing $\mathbf{X}_a$ as the Cholesky decomposition of $\rho^2 \mathbf{I} - \mathbf{X}^T \mathbf{X}$. The matrix $\mathbf{Y}_1$ is arbitrary, and is used as a starting point, although careful choice might be desirable. Koschat and Swayne's algorithm for finding an $\mathbf{A}$ which minimizes R^2 is as follows:

1. Set the starting matrices $\mathbf{X}^*, \mathbf{Y}_1^*$.
2. For $i = 1, 2, \ldots$, find $\mathbf{H}_i$, the orthogonal rotation that minimizes

$$R^2 = \text{tr}(\mathbf{Y}_i^* - \mathbf{X}^* \mathbf{H}_i) \mathbf{W}_p^2 (\mathbf{Y}_i^* - \mathbf{X}^* \mathbf{H}_i)^T,$$

   using results for the unweighted case, since $\mathbf{X}^{*T} \mathbf{X}^* = \rho^2 \mathbf{I}$. Stop if convergence has been achieved.
3. Compute the updated $\mathbf{Y}^*$ matrix

$$\mathbf{Y}_{i+1}^* = \begin{bmatrix} \mathbf{Y} \\ \mathbf{X}_a \mathbf{H}_i \end{bmatrix}$$

4. Go to 2.

Koschat and Swayne show that the values of R^2 form a nonincreasing sequence and hence converge, and also that $\mathbf{H}_i$ converges if there are only finitely many extrema or saddle points for R^2.

5.4.2 More than two configurations

Instead of two configurations to be matched, suppose there are m configurations that need to be matched simultaneously. Procrustes analysis can be modified to allow for this. Let the configurations be given by matrices $\mathbf{X}_i$ ($i = 1, \ldots, m$), and let $\mathbf{A}_i$ be the orthogonal rotation applied to the ith configuration. Then

$$R^2 = \sum_{i<j} \text{tr}(\mathbf{X}_i\mathbf{A}_i - \mathbf{X}_j\mathbf{A}_j)^T(\mathbf{X}_i\mathbf{A}_i - \mathbf{X}_j\mathbf{A}_j)$$

needs to be minimized.

Kristof and Wingersky (1971) and Gower (1975) give a method for solving this generalized Procrustes problem. Firstly the configurations $\mathbf{X}_i$ are centred at the origin, and scaled uniformly so that $\sum_{i=1}^{m} \text{tr}(\mathbf{X}_i\mathbf{X}_i^T) = m$. The configurations $\mathbf{X}_i$ are rotated in turn to $\mathbf{Y}$, the mean matrix, $m^{-1}\sum \mathbf{X}_i$, using the usual two configuration Procrustes rotation. The mean matrix is then updated after every rotation. The iterations will converge to a minimum for R^2. If scaling of the matrices is required a further step in the algorithm is needed. Ten Berge (1977) considered the algorithm of Kristof and Wingersky in detail and suggested a modification to Gower's method. See also ten Berge and Knol (1984).

Peay (1988) gives a good summary of the problem and suggests a method for rotating configurations which maximizes the matching among subspaces of the configurations.

5.4.3 The coefficient of congruence

Tucker (1951) introduced the coefficient of congruence between two vectors $\mathbf{x}$ and $\mathbf{y}$ as

$$\Gamma(\mathbf{x}, \mathbf{y}) = \mathbf{x}^T\mathbf{y}/(\mathbf{x}^T\mathbf{x}.\mathbf{y}^T\mathbf{y})^{\frac{1}{2}}.$$

The maximum value of Γ is unity when $\mathbf{x} = \lambda\mathbf{y}$, λ a positive constant. Instead of using R^2, the sum of the distances between corresponding points in the two configurations, Korth and Tucker (1976) and Brokken (1983) have suggested using the sum of the coefficients of congruence, g, between the corresponding points. This can be written as

$$g = \text{tr}\{[\text{diag}(\mathbf{A}^T\mathbf{X}^T\mathbf{X}\mathbf{A})]^{-\frac{1}{2}}\mathbf{A}^T\mathbf{X}^T\mathbf{Y}[\text{diag}(\mathbf{Y}^T\mathbf{Y})]^{-\frac{1}{2}}\}.$$

The derivatives necessary to maximize g using Newton-Raphson iteration are given by Brokken.

Using Tucker's coefficients of congruence in the matching of matrices is useful in factor analysis (see for example Mardia *et al.* (1979), Chapter 9), where factor loadings given by matrices $\mathbf{X}$ and

$\mathbf{Y}$ are to be compared. Also useful in factor analysis is the use of oblique rotations of factor matrices. This gives rise to the oblique Procrustes problem.

5.4.4 Oblique Procrustes problem

The oblique Procrustes problem is to find a non-orthogonal rotation matrix $\mathbf{A}$ such that

$$R^2 = \mathrm{tr}(\mathbf{Y} - \mathbf{XA})^T(\mathbf{Y} - \mathbf{XA})$$

is a minimum and subject only to $\mathrm{diag}(\mathbf{A}^T\mathbf{A}) = \mathbf{I}$.

Browne (1967) gave a numerical solution using Lagrange multipliers. No constraints are imposed between columns of $\mathbf{A}$ and hence each column can be considered separately. Let $\mathbf{y}$ be a column of $\mathbf{Y}$. A vector $\mathbf{a}$ has to be found such that

$$(\mathbf{y} - \mathbf{Xa})^T(\mathbf{y} - \mathbf{Xa})$$

is a minimum subject to $\mathbf{a}^T\mathbf{a} = 1$.

Let

$$g = (\mathbf{y} - \mathbf{Xa})^T(\mathbf{y} - \mathbf{Xa}) - \mu(\mathbf{a}^T\mathbf{a} - 1),$$

where μ is a Lagrange multiplier. Differentiating with respect to $\mathbf{a}$ and setting equal to $\mathbf{0}$ gives

$$\mathbf{X}^T\mathbf{Xa} - \mu\mathbf{a} = \mathbf{X}^T\mathbf{y}. \tag{5.6}$$

This equation can be simplified by using the spectral decomposition of $\mathbf{X}^T\mathbf{X}$ $(= \mathbf{U}\boldsymbol{\Lambda}\mathbf{U}^T$ say$)$.

Equation (5.6) becomes

$$\mathbf{U}\boldsymbol{\Lambda}\mathbf{U}^T\mathbf{a} - \mu\mathbf{a} = \mathbf{X}^T\mathbf{y}.$$

Pre-multiply by $\mathbf{U}^T$, let $\mathbf{U}^T\mathbf{a} = \mathbf{b}$ and $\mathbf{U}^T\mathbf{X}^T\mathbf{y} = \mathbf{w}$, then

$$\boldsymbol{\Lambda}\mathbf{b} - \mu\mathbf{b} = \mathbf{w}. \tag{5.7}$$

Now $\mathbf{a}^T\mathbf{a} = 1$ and hence the equation now to be solved can be written

$$b_i = \frac{w_i}{\lambda_i - \mu}, \qquad \sum b_i^2 = 1,$$

and hence the roots of $z(\mu) = 0$ are required, where

$$z(\mu) = \sum \frac{w_i^2}{(\lambda_i - \mu)^2} - 1$$

giving the stationary points of R^2.

Browne goes on to show that the minimum value of R^2 corresponds to the smallest real root of $z(\mu)$. He uses the Newton-Raphson method to solve for the roots.

Cramer (1974) pointed out that if $\mathbf{X}^T\mathbf{y}$ is orthogonal to the eigenvector corresponding to the smallest eigenvalue λ_p then there may be a solution to (5.7) which is not a solution of $z(\mu) = 0$. Ten Berge and Nevels (1977) give a solution which covers this case.

5.4.5 Perturbation analysis

Sibson (1979) investigated the distribution of R^2 when a configuration matrix $\mathbf{X}$ is perturbed with random errors added to its elements, $\mathbf{X} + \epsilon\mathbf{Z}$, and then matched back to the original $\mathbf{X}$.

Let $\mathbf{X}^T\mathbf{X}$ have eigenvalues $\lambda_1 > \ldots > \lambda_n > 0$ and corresponding eigenvectors $\mathbf{v}_1, \ldots, \mathbf{v}_n$. Sibson shows that if dilation is not included

$$R^2 = \frac{1}{2}\epsilon^2 \left\{ \sum_{i=1}^{p}\sum_{j=1}^{p} \frac{\mathbf{v}_i^T(\mathbf{X}^T\mathbf{Z} + \mathbf{Z}^T\mathbf{X})\mathbf{v}_j)^2}{\lambda_i + \lambda_j} \right.$$

$$\left. + 2\sum_{i=p+1}^{n-1} \mathbf{v}_i^T\mathbf{Z}\mathbf{Z}^T\mathbf{v}_i \right\}. \tag{5.8}$$

Thus it can be shown that if the elements of $\mathbf{Z}$ are independent $N(0,1)$ random variables, then approximately

$$R^2 \sim \epsilon^2 \chi^2_{np-\frac{1}{2}p(p+1)}.$$

If dilation of a configuration is included the term

$$-2\frac{(\mathrm{tr}\mathbf{X}^T\mathbf{Z})^2}{\mathrm{tr}\mathbf{X}^T\mathbf{X}}$$

has to be included in (5.8) and approximately

$$R^2 \sim \epsilon^2 \chi^2_{np-\frac{1}{2}p(p+1)-1}.$$

Langron and Collins (1985) extend the work of Sibson to the generalized case of several configuration matrices. They consider the two configuration situation with errors in both configuration matrices and show

$$R^2 \sim 2\epsilon^2 \chi^2_{np-\frac{1}{2}p(p+1)}, \qquad R^2 \sim 2\epsilon^2 \chi^2_{np-\frac{1}{2}p(p+1)-1},$$

approximately for the cases of no dilation allowed and dilation allowed respectively. They generalize this to the situation of m configurations. They also show how an ANOVA can be carried out to investigate the significance of the different parts of the Procrustes analysis, translation, rotation/reflection and dilation. The reader is referred to their paper for further details.

CHAPTER 6

Monkeys, aeroplanes, yoghurts and bees

6.1 Introduction

Metric and nonmetric multidimensional scaling is currently being used for data analysis in a multitude of disciplines. Some relatively recent examples are: biometrics – Lawson and Ogg (1989); counselling psychology – Fitzgerald and Hubert (1987); ecology – Tong (1989); ergonomics – Coury (1987); forestry – Smith and Iles (1988); lexicography – Tijssen and Van Raan (1989); marketing – Büyükkurt and Büyükkurt (1990); and tourism – Fenton and Pearce (1988).

In this chapter four applications of multidimensional scaling are reported. The examples come from the areas of animal behaviour, defence, food science and biological cybernetics.

6.2 Monkeys

Corrandino (1990) used MDS to study the proximity structure in a colony of Japanese monkeys. Observations were made on a social group of 14 Japanese monkeys over a period of a year. The fourteen monkeys are named and described in Table 6.1.

Proximity relations every 60 seconds were observed. If two monkeys were within 1.5 m of each other and were tolerating each other, then they were said to be "close". Dissimilarities were calculated for each pair of monkeys based on the amount of time the pair were in proximity to one another. The dissimilarities were then subjected to nonmetric MDS, proximities in the breeding season and non-breeding season being treated separately.

Figure 6.1(i) shows the two dimensional configuration for the non-breeding season, and Figure 6.1(ii) for the breeding season. The two configurations have been aligned using Procrustes analysis. The stress was 28% for the non-breeding season and 29% for

(i)

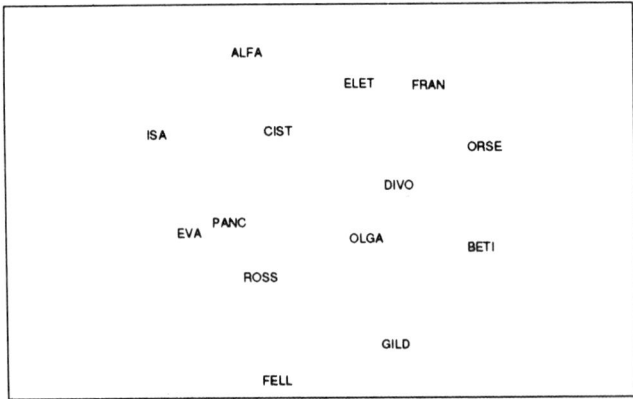

(ii)

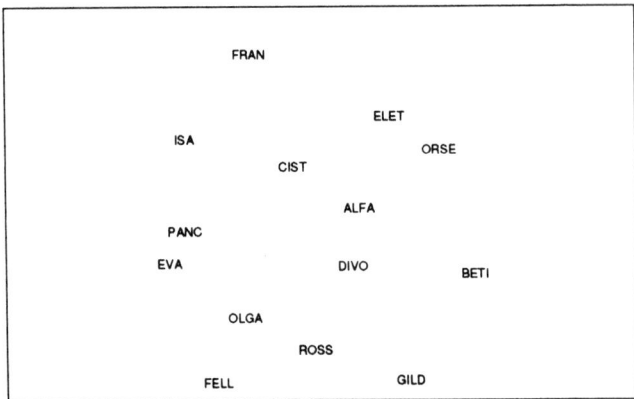

Figure 6.1 *Two dimensional configurations obtained from nonmetric MDS of the monkey data: (i) non-breeding season; (ii) breeding season.*

the breeding season. These values are very high indicating a poor fit. The latter value agrees with that of Corradino but not the first. Although the stress is high some interpretation can be placed on the configurations. Firstly, the three infant/juveniles (FELL, ORSE, EVA) are on the edges of the configurations. There is a line of males (BETI, DIVO, CIST, ALFA) through the females

Table 6.1 *The colony of Japanese monkeys.*

Monkey	age/sex	Monkey	age/sex
Alfa (ALFA)	Adult male	Olga (OLGA)	Adult female
Francesca (FRAN)	Adult female	Orsetta (ORSE)	Inf/juv female
Fello (FELL)	Inf/juv male	Rossella (ROSS)	Adult female
Pancia (PANC)	Adult female	Divo (DIVO)	Subadult male
Isa (ISA)	Adult female	Cisto (CIST)	Subadult male
Gilda (GILD)	Adolescent female	Elettra (ELET)	Adult female
Betino (BETI)	Subadult male	Eva (EVA)	Inf/juv female

(GILD, ROSS, OLGA, PANC, ISA, FRAN, ELET). In the non-breeding season the only fully adult male (ALFA) is on the edge of the configuration while in the breeding season he is right at the centre of things!

6.3 Aeroplanes

Polzella and Reid (1989) used nonmetric MDS on performance data from simulated air combat manoeuvring, collected by Kelly *et al.* (1979). Data were collected for experienced and novice pilots. The variables measured included aircraft system variables, engagement outcomes and events, air combat performance variables, and automatically recorded aircraft variables, e.g. position, altitude. Polzella and Reid used the correlation matrix for thirteen variables measuring pilot performance as similarities for nonmetric scaling, using the SPSS program ALSCAL to perform the analysis.

Figure 6.2(i) shows their two dimensional output for expert pilots, and Figure 6.2(ii) for novice pilots. Stress for the two cases was 6% and 8% respectively.

Their conclusions were that the cluster of variables on the left of Figure 6.2(i) are all energy related indicating that the expert "pilots' performance" was characterized by frequent throttle activity. The cluster of variables on the right are related to air combat manoeuvrability indicating that mission success was associated primarily with offensive and defensive manoeuvrability. The configuration for the novice pilots is markedly different from that for the expert pilots. "Gun kill" was isolated from the other variables indicating mission success was not related to efficient energy management or skilful flying. The variable "fuel flow" being close to

(i)

(ii)

(iii)

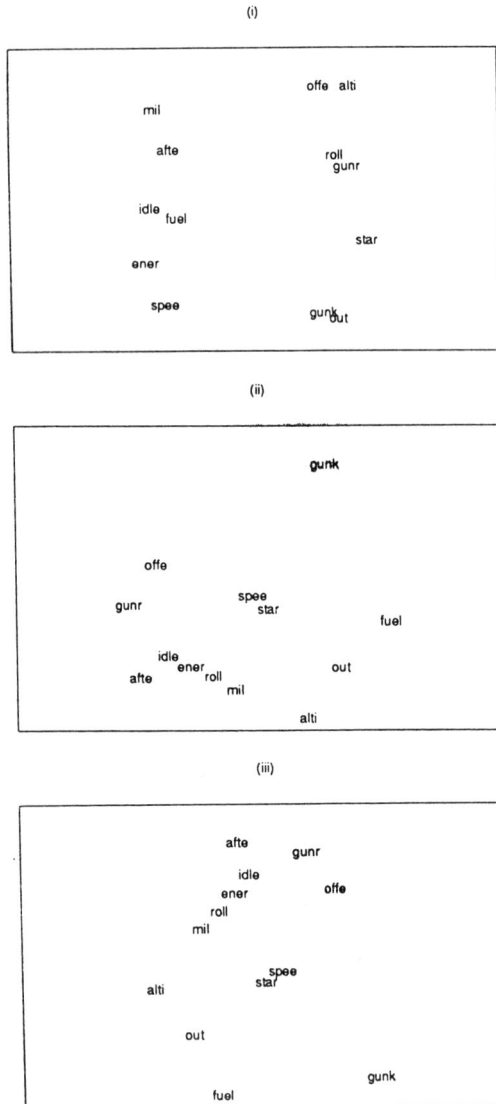

Figure 6.2 *MDS configurations for pilots: (i) experienced pilots; (ii) novice pilots; (iii) the configuration for novice pilots aligned with that for the expert pilots using Procrustes analysis.*

the variables "out of view" and "altitude rate" suggests defensive action for novice pilots requires excessive fuel consumption compared to the expert pilots. Also with "offense", "gun range" and "roll rate" within a cluster of energy related variables, novice pilots made more use of throttle activity for offensive flying than did the expert pilots.

A Procrustes analysis was not carried out by Polzella and Reid in order to allign the two configurations. When this is done the Procrustes statistic has the value is 0.88 indicating a substantial difference in the configurations. Figure 6.2(iii) shows the resulting rotated, reflected, dilated configuration for the novice pilots matched to that of the expert pilots

6.4 Yoghurts

Poste and Patterson (1988) carried out metric and nonmetric MDS analyses on yoghurts. Twelve commercially available yoghurts (four firm, eight Swiss style) were evaluated by ten judges on nine variables. Strawberry yoghurts were presented in pairs to the panelists who were asked to evaluate how similar the two samples were on a 15 cm descriptive line scale. Panelists were asked about the following attributes: colour, amount of fruit present, flavour, sweetness, acidity, lumpiness, graininess, set viscosity, aftertaste. Numerical scores were obtained from the scales, which were then used to compute a correlation matrix for the nine attributes. Metric and nonmetric scaling were used. Unfortunately their results give the stress in two dimensions as 31% and in three dimensions as 22%. From Figure 3.5 of Chapter 3 it can be seen that for twelve points the mean stress for a random ranking of dissimilarities is about 22% for two dimensions and about 12% for three dimensions. A mistake is indicated somewhere in their analysis. Also the configuration they obtained has ten points forming a circle with two points enclosed within the circle. All the points are straining to be as far away from each other as possible, but subject to the normalizing constraint. This could have happened if similarities were accidently used as dissimilarities.

Included in the paper are mean scores for the nine variables for each of the yoghurts. Measuring dissimilarity by Euclidean distance, a configuration of points representing the yoghurts was found using nonmetric scaling. The stress was 13% which is rather high for only twelve points. The configuration is shown in Figure

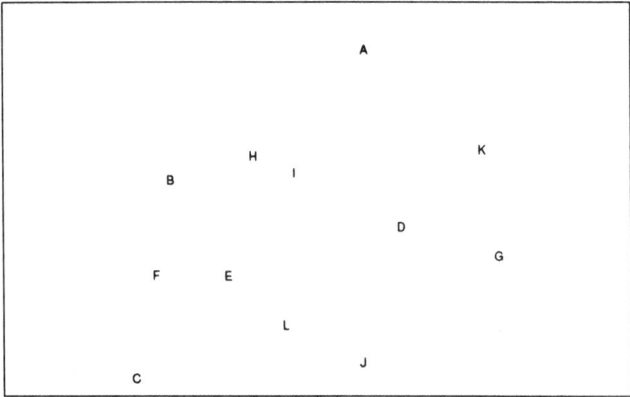

Figure 6.3 *Nonmetric MDS configuration for yoghurts. Swiss style: A, B, C, E, F, H, I, L. Firm: D, G, J, K.*

6.3. The firm yoghurts, D,G,J,K, are towards the lower right hand corner of the configuration. The Swiss style yoghurts can be "imaginatively" ordered by projecting the points representing them onto a line at 45° to the configuration. The order is A I H B E L F C. This agrees well with the configuration obtained by Poste and Patterson when they used metric scaling on the data.

6.5 Bees

Bees have been used in many experiments designed to study their colour vision. Bees' colour perception is investigated by analysing frequency data of the choice made between various colours when they search for food. For example a bee can first be trained to the colour green by always supplying it with food from a green container. Later the bee has to seek food having a choice of blue and green containers to visit, but where the food is only placed in the green container. The bee has to search for food several times and the frequency with which it visits a green container first is recorded.

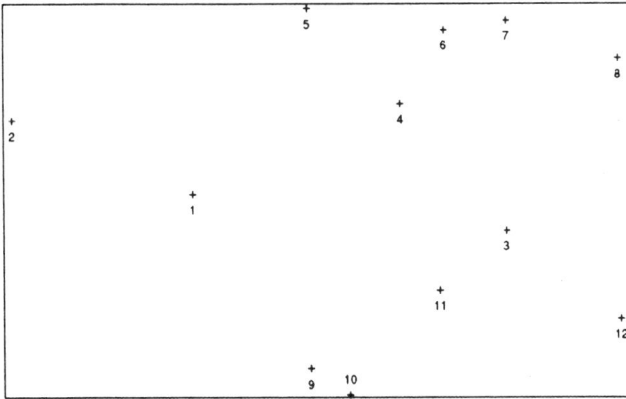

Figure 6.4 *Two dimensional MDS configuration for colour stimuli for bees.*

Backhaus *et al.* (1987) report on an experiment where multidimensional scaling was used on colour similarity data for bees. Firstly each bee was trained to one of twelve colour stimuli by rewarding it with food. Then each bee was tested by giving it a choice of colour in its search for food. Some bees were given a choice of two colours, others a choice of all twelve colours. Multiple choice data were converted to dual choice data as follows.

Let $_t f_r$ be the frequency with which bees trained to colour stimulus t mistakenly choose colour stimulus r. Then $_t \hat{p}_{rs} = {_t f_r}/({_t f_r} + {_t f_s})$ is the proportion of times colour stimulus r was judged more similar to the training colour t than colour stimulus s was so judged. As dual choice proportions obtained from the multiple choice tests were not significantly different from those for dual choice tests, the multiple choice data were included in the MDS analysis.

Define $_t z_{rs}$ by $_t z_{rs} = \Phi^{-1}(_t \hat{p}_{rs})$, the inverse of the standard normal distribution function, for which the approximation can be made

$$_t z_{rs} = \frac{1}{(8\pi)^{\frac{1}{2}}} \ln \left\{ \frac{_t \hat{p}_{rs}}{(1 - {_t \hat{p}_{rs}})} \right\}.$$

The dissimilarities between the colours $\{\delta_{rs}\}$ are assumed to satisfy

$$_t z_{rs} = \delta_{tr} - \delta_{ts}.$$

Let $\delta_{rs} = h_{rs} + c$, where c is an unknown additive constant. Then h_{rs} can be estimated by

$$h_{rs} = \tfrac{1}{2}(_r h_{.s} + .z_{r.} + _s z_{r.} + .z_{.s}),$$

Backhaus *et al.* subjected the derived dissimilarities to metric and nonmetric scaling in a thorough investigation. The Minkowski metric was used with various values of the exponent λ. The city block metric ($\lambda = 1$) and the dominance metric ($\lambda \to \infty$) gave the smallest stress values. Euclidean distance ($\lambda = 2$) gave the highest stress values. This was true for two, three and four dimensinal solutions.

Figure 6.4 shows their two dimensional solution using the city block metric. The twelve colour stimuli are: 1. aluminium + foil; 2. grey; 3. BV1; 4. BV2; 5. BV3; 6. BV3 + foil; 7. BV3 + double foil; 8. BG18; 9. GR4; 10. GR4 + foil; 11. GR4 + double foil; 12. VG6. The stimuli 3 to 12 are varying shades of blue-green with or without foils which decreased reflection. In the configuration the aluminium and grey stimuli are well to the left, the more blue than green stimuli are towards the top right, and the more green than blue are towards the bottom right. From all their analyses, Backhaus *et al.* conclude that bees main perceptual parameters are hue (blue/green) and saturation (UV/blue-greenness), and that brightness is ignored by bees.

CHAPTER 7

Unfolding

7.1 Introduction

Models for "unfolding" can be categorized into unidimensional or multidimensional models and also metric or nonmetric models. Coombs (1950) first introduced unfolding as the following unidimensional nonmetric model. Suppose n judges consider a set of m objects (stimuli) and individually rank them. Coombs suggested that the judges and objects could be represented by points on a straight line (scale), where for each judge, the rank order of the distances from his point to the points representing the m objects is the same as his original rank ordering of the objects. For example, suppose there are two judges (1,2) and five essays (A,B,C,D,E) to be judged and ranked in order to allocate prizes. Suppose the judges rank the essays as follows

	1st	2nd	3rd	4th	5th
Judge 1	B	C	A	E	D
Judge 2	A	B	C	E	D

Then the seven points in Figure 7.1 (top line) represent the judges and the five essays. It can be seen that the distances from judge 1 to the five essays have the same ranking as his original ranking of the essays. Similarly for judge 2. The term "unfolding" was coined since for each judge the line can be folded together at the judge's point and his original rankings are observed. These unfoldings can be seen in Figure 7.1 for the two judges. Alternatively, looking at the situation in reverse the judges' rankings when placed on a line can be "unfolded" to obtain the "common" ordering of the objects.

This unidimensional model can be extended to p dimensions simply by placing the $m + n$ points for judges and objects in a p dimensional space and then using distances, Euclidean or otherwise,

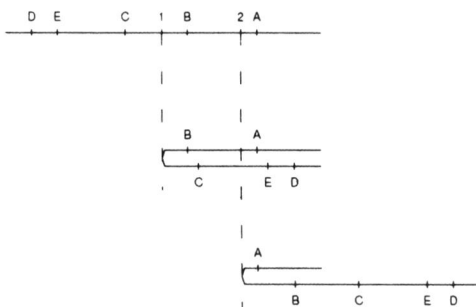

Figure 7.1 *Five essays (A,B,C,D,E) ranked by two judges (1,2), together with their unfoldings.*

in this p dimensional space to determine the rankings for the individual judges. For the "folding" of the space, Coombs used the simile of picking up a handkerchief (p dimensional!) at a judge's point, and letting the ends fall together to determine the rankings for that judge. The metric unfolding model is very similar to the nonmetric model but dissimilarities replace the rankings by the judges, and distances from a judge's point in the folding space to the objects are to match the original dissimilarities. Of course the matching of distances to dissimilarities or ranks to ranks can never be guaranteed and so compromises have to be made.

7.2 Nonmetric unidimensional unfolding

Coombs (1950, 1964) introduced the J scale and I scale. The line upon which points are placed for the n judges or individuals together with the m stimuli is called the J scale. Each individual's preference ordering is called an I scale. Consider just four stimuli A,B,C and D. Figure 7.2 shows these on a J scale.

Also shown are all the midpoints between pairs of stimuli where for instance AB denotes the midpoint between A and B. The J scale can be seen to be split into seven intervals. Any judge represented by a point in a particular interval will have the same I scale as any other judge in that interval. For example, a point in interval I_5 has the preference ordering CBDA. Table 7.1 gives the preference ordering for the seven intervals.

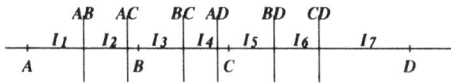

Figure 7.2 *Four stimuli (A,B,C,D) placed on a J scale together with intervals for I scales.*

Table 7.1 *Preference orderings for the J scale.*

Interval:	I_1	I_2	I_3	I_4	I_5	I_6	I_7
Ordering:	ABCD	BACD	BCAD	CBAD	CBDA	CDBA	DCBA

Of course, not all preference orderings can be accommodated. For instance, DABC in this example is impossible to achieve. For a given J scale it is easy to generate the possible I scales. However, the more challenging task is to generate the J scale from a set of I scales.

When a J scale is folded to produce an I scale, the I scale must end with either of the two end stimuli of the J scale. Hence the end stimuli for the J scale will be known by simply looking at the end stimuli for all the I scales. There will be only two I scales starting and ending with the two end points of J, and these two I scales will be mirror images of each other. There will be no other I scales which are mirror images of each other. The rank order of the J scale can be taken as the rank order of either of the mirror image I scales. Next, the order of the midpoints as in Figure 7.1 needs to be determined. This is done by determining the order of the I scales. Note that in Table 7.1, as the intervals are read from left to right, an adjacent pair of stimuli are interchanged at each step. For example, the first interval corresponds to the ordering ABCD, whereupon interchanging A and B gives the ordering BACD of the second interval. Then interchanging A and C gives the third, etc. The interchanging of a pair of stimuli corresponds to the crossing of the midpoint of that pair. Hence for given preference orderings, to order the midpoints, the I scales are ordered accordingly, starting with one of the mirror image I scales and ending with the other. In practice further geometrical considerations have to be taken into

account and the reader is referred to Coombs (1964) for further
details, and also for details of applications of the technique to sets
of psychological data.

The main problem with this unfolding technique is that for a
given set of I scales it is unlikely that a single J scale can be
found. Hettmansperger and Thomas (1973) attempt to overcome
this problem by using a probability model. For a given number of
stimuli, the probabilities of the various possible I scales for a given
J scale are taken to be constant, i.e. $P(I_i|J_j) = c$. Then $P(I_i) = \sum_j P(I_i|J_j)P(J_j) = c\sum_j P(J_j)$ is used to form the likelihood for
a sample of N I scales. From the likelihood, the various $P(J_j)$ are
estimated.

Zinnes and Griggs (1974) use a different probability model. They
assume that each individual, r, is placed independently on a point
x_r on the J scale, where $x_r \sim N(\mu_r, \sigma_r^2)$. The jth stimulus is placed
independently at the point y_j, where $y_j \sim N(\xi_j, \nu_j^2)$. Then the
probability that individual r prefers stimulus i to stimulus j is
$p_{ij} = \Pr(|x_r - y_i| < |x_r - y_j|)$. For the case $\sigma_r^2 = \sigma_i^2 = \frac{1}{2}$ for all i,
Zinnes and Griggs show that

$$p_{ij} = 1 - \Phi(a_{ij}) - \Phi(b_{ij}) + 2\Phi(a_{ij})\Phi(b_{ij}),$$

where

$$a_{ij} = (2\mu_r - \xi_i - \xi_j)/\sqrt{3}, \quad b_{ij} = \xi_i - \xi_j,$$

with similar results under different assumptions regarding the vari-
ances. Data collected from individuals are then used to find maximum likelihood estimates of $\{\xi_i\}$ and $\{\mu_r\}$ in order to draw up
a J scale. The data for an individual can be in the form of preference data for pairs, or can be extracted in this form from the
individual's preference ranking.

7.3 Nonmetric multidimensional unfolding

Bennett and Hays (1960) and Hays and Bennett (1961) generalized
Coomb's unidimensional unfolding model to several dimensions.
Their work is also reported in Coombs (1964). Great detail is not
gone into here since the theory is similar to that for the unidimensional case, except that the geometrical structure is much more
complicated. Some of their results are summarized.

Consider points representing individuals and stimuli placed in a
space of p dimensions. The locus of points, equidistant from stimuli

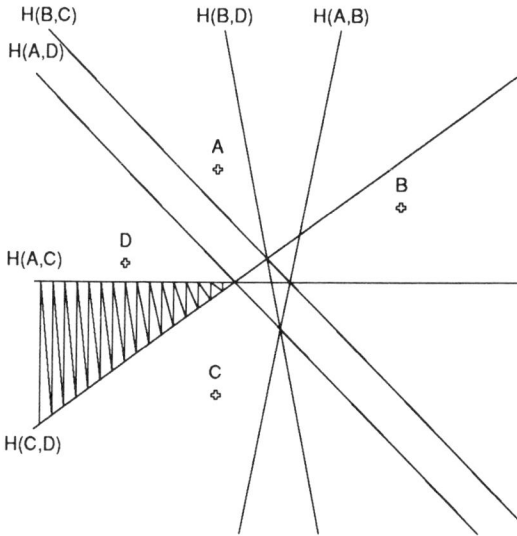

Figure 7.3 *Isotonic regions for four stimuli in a two dimensional space.*

A and B, is a hyperplane, $H(A, B)$ of dimension $p - 1$. Similarly, the locus of points equidistant from m stimuli (assuming none of the stimuli are coplanar) is a hyperplane, $H(A, B, ...)$ of dimension $p - m + 1$. The hyperplane $H(A, B)$ splits the space into two half spaces, where an individual placed in one half space prefers A to B and if placed in the other, prefers B to A. Bennett and Hays call these half spaces or zones, the isotonic regions AB and BA, indicating the preferred orderings. The space can be divided up into isotonic regions by the hyperplanes defined by each pair of stimuli. Figure 7.3 shows the case for $p = 2$ and $n = 4$. The hyperplanes are straight lines. The isotonic regions are labelled according to the preferred order for the points in a particular isotonic region. For example all points in the shaded isotonic region have the preffered ordering D,C,A,B.

As in the unidimensional case, certain preferred orderings cannot occur. This will happen when there are more than $p + 2$ stimuli in a p dimensional space. Again, it is a relatively easy task to divide a p dimensional space, with stimuli in fixed positions (J scale), into the various isotonic regions, although for high dimensional space the lack of a graphical illustration will detract from the interpretation. The more difficult task is to construct a configuration of points in p dimensions, representing the stimuli from a set of experimentally obtained I scales.

Bennett and Hays tackle the problem of determining the required dimension of the space in three ways. It is of course always possible to place the points in a space of $n - 1$ dimensions to recover all possible rankings of the n stimuli. However, to be useful, a space of a much lower dimension is required. Their first method for determining dimension is based on the bounding of isotronic regions and can only be used for $p = 1$ or 2. Their second method is based on the number of isotronic regions that can be generated from m stimuli in p dimensions, $c(m, p)$ say. They give the recurrence relation

$$c(m, p) = c(m - 1, p) + (m - 1)c(m - 1, p - 1)$$

and a corresponding table of values of $c(m, p)$ for $m = 1(1)20$, $p = 1(1)5$. It is not surprising that $c(m, p)$ is much less than the total number of possible rankings, $m!$, for reasonably large m and small dimension p. For example, for $m = 9$, $p = 3$, $m! = 362\,880$ and $c(m, p) = 5119$. From experimental data, the total number of rankings of the m stimuli by the individuals is obtained and then dimensionality is assessed by comparing this with values of $c(m, p)$.

Bennett and Hays' third method for determining dimension is based on the result that the minimum dimension of the space in which a complete solution may be realized must be one less than the number of elements in the largest transposition group of stimuli present in the experimental data. The reader is referred to the papers by Bennett and Hays for further details.

McElwain and Keats (1961) considered the case of four stimuli in two dimensions in detail. They looked at the I scales generated by all possible geometric configurations of stimuli and then were able to characterize a set of I scales to determine the type of geometrical configuration required. For more than four stimuli or more than two dimensions, this would appear to be an impossible task.

Davidson (1972, 1973) gives further geometric results for nonmetric multidimensional unfolding. He derives necessary and sufficient conditions for a configuration of points representing stimuli to give rise to a particular set of I scales. Also he gives the necessary and sufficient geometric constraints that determine the subset of pairs of orders and opposites of the stimuli, contained in the particular set of I scales.

For a probabilistic model, Zinnes and Griggs (1974) extend their unidimensional model by using vector random variables $\mathbf{x}_r$ and $\mathbf{y}_j$, where $\mathbf{x}_r \sim \text{MVN}(\boldsymbol{\mu}_r, \sigma^2 \mathbf{I})$, $\mathbf{y}_j \sim \text{MVN}(\boldsymbol{\xi}_j, \nu^2 \mathbf{I})$. Then p_{ij} is given by

$$p_{ij} = \text{P}(F''_{ij} \leq 1),$$

where F'' is the doubly non-central F distribution, $F(\nu_i, \nu_j, \lambda_i, \lambda_j)$, with

$$\nu_i = \nu_j = p$$
$$\lambda_i = (\boldsymbol{\mu}_r - \boldsymbol{\mu}_i)^T (\boldsymbol{\mu}_r - \boldsymbol{\mu}_i)$$
$$\mu_j = (\boldsymbol{\mu}_r - \boldsymbol{\mu}_j)^T (\boldsymbol{\mu}_r - \boldsymbol{\mu}_j).$$

A normal approximation is used for F'' and then the method proceeds as for the unidimensional case.

7.4 Metric multidimensional unfolding

The case of metric unidimensional unfolding will be subsumed in the multidimensional case. Coombs and Kao (1960) and Coombs (1964) started to look at a metric method for unfolding by using a principal components analysis on the correlation matrix obtained from the correlations between pairs of I scales. Ross and Cliff (1964) took the method further. Schönemann (1970) found an algebraic solution for metric unfolding.

As before let there be n individuals or judges, and suppose the rth individual produces dissimilarities $\{\delta_{ri}\}$ for m stimuli. Suppose $m + n$ points are placed in a p dimensional Euclidean space where each individual and each stimulus is represented by one of the points. Let the coordinates of the points representing the individuals be $\mathbf{x}_r$ $(r = 1, \ldots, n)$ and the coordinates of the points representing the stimuli be $\mathbf{y}_i$ $(i = 1, \ldots, m)$. Let the distance between the points representing the rth individual and the ith stimulus be d_{ri}. The metric unfolding problem is to find a configuration such that the distances $\{d_{ri}\}$ best represent the dissimilarities $\{\delta_{ri}\}$.

Schönemann (1970) gave an algorithm to find $\{\mathbf{x}_r\}$, $\{\mathbf{y}_i\}$ from the distances $\{d_{ri}\}$. Gold (1973) clarified Schönemann's work; the following is a brief summary.

Let $\mathbf{X} = [\mathbf{x}_1, \ldots, \mathbf{x}_n]^T$, $\mathbf{Y} = [\mathbf{y}_1, \ldots, \mathbf{y}_m]^T$. Let the matrix of squared distances between the points representing the individuals and the points representing the objects be $D(\mathbf{X}, \mathbf{Y})$. Hence

$$[D(\mathbf{X}, \mathbf{Y})]_{ri} = (\mathbf{x}_r - \mathbf{y}_i)^T(\mathbf{x}_r - \mathbf{y}_i).$$

Let the matrix of squared dissimilarities be $\mathbf{D} = [\delta_{ri}^2]$. The metric unfolding problem is to find $(\mathbf{X}, \mathbf{Y})$ such that $D(\mathbf{X}, \mathbf{Y}) = \mathbf{D}$.

The matrices $\mathbf{D}$ and $D(\mathbf{X}, \mathbf{Y})$ are now doubly centred to give $\mathbf{C} = \mathbf{HDH}$, and $\mathbf{C}(\mathbf{X}, \mathbf{Y}) = \mathbf{H}D(\mathbf{X}, \mathbf{Y})\mathbf{H}$, where $\mathbf{H}$ is the centring matrix.

Then the unfolding problem can be rewritten as

$$\mathbf{C}(\mathbf{X}, \mathbf{Y}) = \mathbf{C} \tag{7.1}$$

$$D(\mathbf{X}, \mathbf{Y})_{r.} = \mathbf{D}_{r.} \quad (r = 1, \ldots, n) \tag{7.2}$$

$$D(\mathbf{X}, \mathbf{Y})_{.i} = \mathbf{D}_{.i} \quad (i = 1, \ldots, m). \tag{7.3}$$

The matrices $(\mathbf{X}, \mathbf{Y})$ satisfying these equations are called an unfolding. Schönemann's algorithm requires two steps. Step 1 is to find those unfoldings $(\mathbf{X}, \mathbf{Y})$ which satisfy (7.1), Step 2 is then to find which unfoldings of Step 1 satisfy (7.2) and (7.3). For further details see Schönemann (1970) and Gold (1973).

A more useful approach is the introduction of a loss function as in Greenacre and Browne (1986). They proposed an efficient alternating least squares algorithm for metric unfolding. It is the one used in this book to analyse example data and the program for it is included on the accompanying computer diskette. A brief description is given. The algorithm uses squared Euclidean distances $\{d_{ij}^2\}$ to approximate to the squared dissimilarities $\{\delta_{ij}^2\}$. This leads to a simplification over the use of non-squared distances and dissimilarities. Using the previous notation, the model which incorporates residuals $\{\epsilon_{ri}\}$ is

$$\delta_{ri}^2 = d_{ri}^2 + \epsilon_{ri},$$

or

$$\delta_{ri}^2 = (\mathbf{x}_r - \mathbf{y}_i)^T(\mathbf{x}_r - \mathbf{y}_i) + \epsilon_{ri}.$$

An unfolding $(\mathbf{X}, \mathbf{Y})$ is then found that minimizes

$$\sum_r \sum_i \epsilon_{ri}^2 = \text{tr}(\mathbf{RR}^T),$$

where $[\mathbf{R}]_{ri} = \epsilon_{ri}$.

Following Greenacre and Browne, let

$$f(\mathbf{X}, \mathbf{Y}; \mathbf{D}^{(2)}) = \sum_{r=1}^{n} \sum_{i=1}^{m} \{\delta_{ri}^2 - (\mathbf{x}_r - \mathbf{y}_i)^T (\mathbf{x}_r - \mathbf{y}_i)\}^2 \qquad (7.4)$$

where $[\mathbf{D}^{(2)}]_{ri} = \delta_{ri}^2$.
Then

$$\frac{\partial f}{\partial \mathbf{x}_r} = 4 \sum_{i=1}^{m} \{\delta_{ri}^2 - (\mathbf{x}_r - \mathbf{y}_i)^T (\mathbf{x}_r - \mathbf{y}_i)\}(\mathbf{x}_r - \mathbf{y}_i),$$

and equating to $\mathbf{0}$ gives

$$\sum_{i=1}^{m} \{\delta_{ri}^2 - (\mathbf{x}_r - \mathbf{y}_i)^T (\mathbf{x}_r - \mathbf{y}_i)\}\mathbf{y}_i = \sum_{i=1}^{m} \{\delta_{ri}^2 - (\mathbf{x}_r - \mathbf{y}_i)^T (\mathbf{x}_r - \mathbf{y}_i)\}\mathbf{x}_r.$$

This can be written as

$$\sum_{i=1}^{m} [\mathbf{R}]_{ri} \mathbf{y}_i = \sum_{i=1}^{m} [\mathbf{R}]_{ri} \mathbf{x}_r.$$

Combining these equations gives

$$\mathbf{RY} = \text{diag}(\mathbf{RJ}^T)\mathbf{X} \qquad (7.5)$$

where $\mathbf{J}$ is an $n \times m$ matrix of ones, and $\text{diag}(\mathbf{M})$ is the diagonal matrix formed from the diagonal of a matrix $\mathbf{M}$. Similarly

$$\mathbf{R}^T \mathbf{X} = \text{diag}(\mathbf{R}^T \mathbf{J})\mathbf{Y}. \qquad (7.6)$$

Equations (7.5) and (7.6) need to be solved numerically to find an unfolding $(\mathbf{X}, \mathbf{Y})$ giving minimum sum of squared residuals.

Greenacre and Browne use an alternating least squares procedure to minimize (7.4). Their iterative scheme first holds $\mathbf{Y}$ fixed and minimizes (7.4) with respect to $\mathbf{X}$, and then holds $\mathbf{X}$ fixed and minimizes (7.4) with respect to $\mathbf{Y}$. Convergence is guaranteed but can be very slow. A brief description of the derivation of the algorithm is given.

Consider $\mathbf{Y}$ fixed, and write $f(\mathbf{X}, \mathbf{Y})$ as $\sum_{r=1}^{n} f_r$, where

$$f_r = \sum_{i=1}^{m} \{\delta_{ri}^2 - (\mathbf{x}_r - \mathbf{y}_i)^T (\mathbf{x}_r - \mathbf{y}_i)\}^2.$$

Minimizing $f(\mathbf{X}, \mathbf{Y})$ with respect to $\mathbf{X}$ for fixed $\mathbf{Y}$ can be done by minimizing each f_r with respect to $\mathbf{x}_r$ separately.

Differentiating f_r with respect to $\mathbf{x}_r$ and setting equal to $\mathbf{0}$, gives

$$\sum_{i=1}^{m}(\delta_{ri}^2 - \mathbf{x}_r^T\mathbf{x}_r - \mathbf{y}_i^T\mathbf{y}_i + 2\mathbf{x}_r^T\mathbf{y}_i)(\mathbf{x}_r - \mathbf{y}_i) = \mathbf{0}. \qquad (7.7)$$

Greenacre and Browne introduce notation similar to

$$\mathbf{d}_r^{(2)} = [\delta_{r1}^2, \ldots, \delta_{rm}^2]^T$$

$$\mathbf{h} = [\mathbf{y}_1^T\mathbf{y}_1, \ldots, \mathbf{y}_m^T\mathbf{y}_m]^T$$

$$\mathbf{w}_r = \mathbf{Y}^T[\mathbf{d}_r^{(2)} - \mathbf{h}]$$

$$c_r = \mathbf{1}^T[\mathbf{d}_r^{(2)} - \mathbf{h}],$$

and then (7.7) can be written as

$$(c_r - m\mathbf{x}_r^T\mathbf{x}_r - 2\mathbf{1}^T\mathbf{Y}\mathbf{x}_r)\mathbf{x}_r = \mathbf{w}_r - \mathbf{Y}^T\mathbf{1}(\mathbf{x}_r^T\mathbf{x}_r) + 2\mathbf{Y}^T\mathbf{Y}\mathbf{x}_r.$$

They argue that although $\mathbf{Y}$ is fixed, origin and orientation can be chosen and so the choice is made to place the centroid of $\mathbf{Y}$ at the origin and to refer $\mathbf{Y}$ to its principal axes. Thus $\mathbf{Y}^T\mathbf{1} = \mathbf{0}$ and $\mathbf{Y}^T\mathbf{Y} = \mathbf{D}_\lambda$, where $\mathbf{D}_\lambda$ is a diagonal matrix of nonnegative numbers, $\lambda_1 \geq \lambda_2 \geq \ldots \geq \lambda_p$. Obviously if $\mathbf{Y}$ is not in principal axes form, it can be made so by a principle coordinates analysis (PCO) as in Chapter 2. Equation (7.7) becomes

$$(c_r - m\mathbf{x}_r^T\mathbf{x}_r)\mathbf{x}_r - 2\mathbf{D}_\lambda\mathbf{x}_r = \mathbf{w}_r,$$

and hence

$$x_{rk} = \frac{w_{rk}}{c_r - m\mathbf{x}^T\mathbf{x} - 2\lambda_k} \qquad (k = 1, \ldots, p). \qquad (7.8)$$

A variable, ϕ_r, is introduced, where

$$\phi_r = c_r - m\mathbf{x}_r^T\mathbf{x}_r = c_r - m\sum_{k=1}^{p} x_{rk}^2. \qquad (7.9)$$

Hence (7.8) becomes

$$x_{rk} = \frac{w_{rk}}{\phi_r - 2\lambda_k} \qquad (k = 1, \ldots, p). \qquad (7.10)$$

Substituting (7.10) back into (7.9) gives

$$0 = \phi_r - c_r + m\sum_{k=1}^{p} \frac{w_{rk}^2}{(\phi_r - 2\lambda_k)^2} = g(\phi_r)$$

Table 7.2 *The 21 nations.*

Nation		Nation	
Brazil	(BRA)	Israel	(ISR)
China	(CHI)	Japan	(JAP)
Congo	(CON)	Mexico	(MEX)
Cuba	(CUB)	Poland	(POL)
Egypt	(EGY)	USSR	(USSR)
UK	(UK)	South Africa	(SA)
Ethiopia	(ETH)	Spain	(SPA)
France	(FRA)	USA	(USA)
Greece	(GRE)	West Germany	(WG)
India	(INDI)	Yugoslavia	(YUG)
Indonesia	(INDO)		

say. The function $g(\phi_r)$ can then be used to find the required stationary points. If $\phi_r^\star$ is a stationary point of $g(\phi_r)$, then substituting $\phi_r^\star$ into (7.10) gives the stationary point $\mathbf{x}_r^\star$. Greenacre and Browne show that the smallest root of $g(\phi_r) = 0$ is actually the required root to give the global minimum of f_r. Thus the minimization problem for this stage has been reduced to one of finding the smallest root of the equation $g(\phi_r) = 0$. This has to be done for each $\mathbf{x}_r$.

The second stage of the procedure is carried out in a similar manner to the first stage, except that $\mathbf{X}$ is fixed this time. Iterations between the two stages are carried out until convergence is reached.

A starting value for $\mathbf{Y}$ needs to be chosen. Greenacre and Browne suggest using the algorithm of Schönemann (1970). The matrix $-\frac{1}{2}\mathbf{D}^{(2)}$ formed from $\mathbf{d}_r^{(2)}$, is doubly centred to give

$$\mathbf{C} = -\tfrac{1}{2}\mathbf{H}\mathbf{D}^{(2)}\mathbf{H}.$$

The singular value decomposition of $\mathbf{C}$ is found, $\mathbf{C} = \mathbf{U}\mathbf{D}_\alpha\mathbf{V}^T$, where $\mathbf{D}_\alpha = \text{diag}(\alpha_1, \ldots, \alpha_p)$. The starting value for $\mathbf{Y}$ is then taken as $\mathbf{Y}_0 = [\alpha_1\mathbf{v}_1, \ldots, \alpha_p\mathbf{v}_p]$.

7.4.1 The rating of nations

Wish *et al.* (1972) report on a study of the ways that people conceive of nations. Students were asked to judge the similarity between pairs of nations. They were each given a subset of all the

Table 7.3 *The variables scored for the nations.*

Variable	Variable
1 Aligned with USA	10 Population satisfied
2 Individualistic	11 Internally united
3 Peaceful	12 Cultural influence
4 Many rights	13 Educated population
5 I like	14 Rich
6 Good	15 Industrialized
7 Similarity to ideal	16 Powerful
8 Can change status	17 Progressing
9 Stable	18 Large

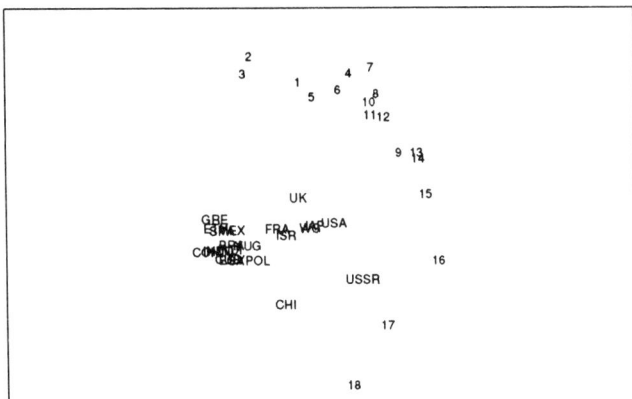

Figure 7.4 *Unfolding analysis of the rating of nations data.*

possible pairs of nations to judge since there were 21 nations in total. The nations are given in Table 7.2. The students were then asked to score each nation on 18 variables on a scale 1 to 9. These variables are given in Table 7.3.

Given in the report are the mean scores of the nations for each of the variables. Wish *et al.* concentrate on using dissimilarities in an individual differences model (INDSCAL) which is described in Chapter 9. Here the mean scores for the nations are subjected to unfolding analysis in a two dimensional space, the nations being treated as "individuals" and the variables as "stimuli". The mean scores were converted to "distances" using the transformation $(9 - \text{mean score})^{\frac{1}{2}}$. Figure 7.4 shows the resulting configuration, which shows some interesting features. The nations split into various groups {UK, USA, Japan, West Germany}, {Greece, Mexico, Ethiopia, Spain}, {Congo, Brazil, Poland, India, Cuba, Indonesia, Yugoslavia, South Africa}, {France, Israel}, {USSR}, and {China}. The 18 variables form a horseshoe and a possible ordering which is approximately the same as their numerical order. Indeed looking at the list of variables, a scale can be envisaged starting from the second variable. The first few variables relate to individuals, the middle to the nation viewed as a population, and the last of the variables to the nation seen as a non-human entity. It is interesting to note the positions of the various nations, realizing of course that the world has progressed since the data were collected in 1968.

CHAPTER 8

Correspondence analysis

8.1 Introduction

Correspondence analysis represents the rows and columns of a data matrix as points in a space of low dimension, and is particularly suited to two-way contingency tables. The method has been discovered and rediscovered several times over the last sixty years and has gone under several different names. Now the most widely accepted name for this particular technique is correspondence analysis, but it is also referred to as "reciprocal averaging" and "dual scaling". Nishisato (1980) and Greenacre (1984) give brief historical accounts of the development. They trace the origins of the method back to Richardson and Kuder (1933), Hirschfeld (1935), Horst (1935), Fisher (1940) and Guttman (1941).

Much of correspondence analysis was developed in France in the 1960s by Benzécri. Benzécri originally called the technique "analyse factorielle des correspondances" but later shortened this to "analyse des correspondances", and hence the English translation. Because correspondence analysis can be related to several other statistical procedures such as canonical correlation analysis, principal components analysis, dual scaling, etc., there are potentially hundreds of references to the subject. Here the method is simply viewed as a metric multidimensional scaling method on the rows and columns of a contingency table or data matrix with non-negative entries.

8.2 Analysis of two-way contingency tables

Suppose data have been collected in the form of an $r \times s$ contingency table. Correspondence analysis finds two vector spaces, one for the rows and one for the columns of the contingency table. These vector spaces give rise to a graphical display of the data. The theory developed will be illustrated by the following example.

Table 8.1 *Malignant melanoma data*

Histological type	Site of tumour		
	Head, neck (h)	Trunk (t)	Extremities (e)
Hutchison's melanotic freckle (H)	22	2	10
Superficial spreading melanoma (S)	16	54	115
Nodular (N)	19	33	73
Interminate (I)	11	17	28

Example

Roberts *et al.* (1981) carried out a study of malignant melanoma, a dangerous type of skin cancer, recording the site of the tumour and also its histological type, for four hundred patients. Results are shown in Table 8.1. These data could be analysed by various more common categorical data methods such as the fitting of log-linear models, see for example Dobson (1983).

These data will be placed in a 4×3 matrix $\mathbf{X}$ and subjected to correspondence analysis. However first the SVD of $\mathbf{X}$ is found for comparison with results from correspondence analysis.

Following Section 1.4.2, the SVD of $\mathbf{X}$ is given by

$$\mathbf{X} = \begin{bmatrix} 0.087 & 0.906 & 0.221 \\ 0.818 & -0.292 & 0.109 \\ 0.526 & 0.215 & 0.187 \\ 0.217 & 0.219 & -0.951 \end{bmatrix} \begin{bmatrix} 156.369 & 0 & 0 \\ 0 & 22.140 & 0 \\ 0 & 0 & 4.083 \end{bmatrix}$$

$$\times \begin{bmatrix} 0.175 & 0.418 & 0.891 \\ 0.982 & -0.144 & -0.125 \\ -0.075 & -0.897 & 0.436 \end{bmatrix}$$

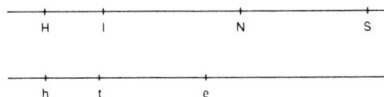

Figure 8.1 *The SVD of the tumour data approximated in one dimension. First space – histological type, second space – site of tumour.*

or equivalently by

$$
\mathbf{X} = 156.369 \begin{bmatrix} 0.015 & 0.036 & 0.078 \\ 0.143 & 0.342 & 0.729 \\ 0.092 & 0.220 & 0.469 \\ 0.038 & 0.091 & 0.194 \end{bmatrix}
$$

$$
+ \ 22.140 \begin{bmatrix} 0.889 & -0.130 & -0.114 \\ -0.287 & 0.042 & 0.037 \\ 0.211 & -0.031 & -0.027 \\ 0.215 & -0.031 & -0.027 \end{bmatrix}
$$

$$
+ \ 4.083 \begin{bmatrix} -0.017 & -0.198 & 0.096 \\ -0.008 & -0.098 & 0.048 \\ -0.014 & -0.167 & 0.081 \\ 0.072 & 0.853 & -0.414 \end{bmatrix}.
$$

Since the first singular value is seven times as large as the second, a one dimensional space could be used to represent the tumour type and also for the site. Figure 8.1 shows a plot of the points corresponding to the first left and right singular vectors. This gives rise to the ordering H,I,N,S for the type of tumour and the ordering h,t,e for the site of the tumour.

Having introduced the data the theory of correspondence analysis is explored in the style of Greenacre (1984).

8.3 The theory of correspondence analysis

Two vector spaces are found, one representing the rows of matrix $\mathbf{X}$ and one representing its columns, in such a way that the rows and columns are similarly treated. This is achieved as follows.

The ith row profile of matrix $\mathbf{X}$ is the ith row of $\mathbf{X}$ standardized so the row sum is unity. The ith row profile is then given a weight r_i, where r_i is the original row sum of $\mathbf{X}$. Similarly the jth column profile of $\mathbf{X}$ is defined as the standardized jth column of $\mathbf{X}$, and given the weight c_j, the jth column sum of $\mathbf{X}$. The matrix of row profiles is given by $\mathbf{D}_r^{-1}\mathbf{X}$ where $\mathbf{D}_r = \text{diag}(r_1, \ldots, r_n)$, and the matrix of column profiles is $\mathbf{D}_c^{-1}\mathbf{X}$, where $\mathbf{D}_c = \text{diag}(c_1, \ldots, c_p)$. The row and column profiles are now represented in vector spaces obtained from the generalized SVD of $\mathbf{X}$.

Let the generalized SVD of $\mathbf{X}$ be given by

$$\mathbf{X} = \mathbf{AD}_\lambda\mathbf{B}^T,$$

where

$$\mathbf{A}^T\mathbf{D}_r^{-1}\mathbf{A} = \mathbf{B}^T\mathbf{D}_c^{-1}\mathbf{B} = \mathbf{I}.$$

The matrix $\mathbf{A}$ is an orthonormal basis for the columns of $\mathbf{X}$, normalized with respect to $\mathbf{D}_r^{-1}$, which allows for the differing row profile weights $\{r_i\}$. Similarly $\mathbf{B}$ is an orthonormal basis for the rows of $\mathbf{X}$, normalized with respect to $\mathbf{D}_c^{-1}$, allowing for the column profile weights $\{c_j\}$.

The row profiles can then be expressed as

$$\mathbf{D}_r^{-1}\mathbf{X} = \mathbf{D}_r^{-1}\mathbf{AD}_\lambda\mathbf{B}^T,$$

with

$$(\mathbf{D}_r^{-1}\mathbf{A})^T\mathbf{D}_r(\mathbf{D}_r^{-1}\mathbf{A}) = \mathbf{B}^T\mathbf{D}_c^{-1}\mathbf{B} = \mathbf{I}.$$

Letting $\mathbf{U} = \mathbf{D}_r^{-1}\mathbf{A}$,

$$\mathbf{D}_r^{-1}\mathbf{X} = \mathbf{UD}_\lambda\mathbf{B}^T, \quad \mathbf{U}^T\mathbf{D}_r\mathbf{U} = \mathbf{B}^T\mathbf{D}_c^{-1}\mathbf{B} = \mathbf{I}. \tag{8.1}$$

Similarly the column profiles can be expressed as

$$\mathbf{D}_c^{-1}\mathbf{X}^T = \mathbf{D}_c^{-1}\mathbf{BD}_\lambda\mathbf{A}^T,$$

with

$$\mathbf{A}^T\mathbf{D}_r^{-1}\mathbf{A} = (\mathbf{D}_c^{-1}\mathbf{B})^T\mathbf{D}_c(\mathbf{D}_c^{-1}\mathbf{B}) = \mathbf{I},$$

and letting $\mathbf{V} = \mathbf{D}_c^{-1}\mathbf{B}$,

$$\mathbf{D}_c^{-1}\mathbf{X}^T = \mathbf{VD}_\lambda\mathbf{A}^T, \quad \mathbf{A}^T\mathbf{D}_r^{-1}\mathbf{A} = \mathbf{V}^T\mathbf{D}_c\mathbf{V} = \mathbf{I}. \tag{8.2}$$

Equation (8.1) shows the row profiles can be represented in the $\mathbf{UD}_\lambda$ space, with $\mathbf{B}$ the rotation matrix which transforms the row profiles to points in the $\mathbf{UD}_\lambda$ space.

For a low dimensional representation of the row profiles, the generalized SVD allows the first k columns of $\mathbf{UD}_\lambda$ to be taken as the best least squares approximating space of dimension k. Similarly

(8.2) shows the column profiles can be represented in the $\mathbf{VD}_\lambda$ space, with $\mathbf{A}$ the necessary rotation matrix.

8.3.1 The cancer example

Without loss of generality the data matrix $\mathbf{X}$ is normalized by dividing each element by the total sum of the elements, i.e. 400. Following the theory just given the relevant matrices are:

$$\mathbf{X} = \begin{bmatrix} 0.055 & 0.005 & 0.025 \\ 0.040 & 0.135 & 0.288 \\ 0.048 & 0.083 & 0.183 \\ 0.028 & 0.043 & 0.070 \end{bmatrix}$$

$$\mathbf{D}_r = \mathrm{diag}[0.085, \quad 0.463, \quad 0.313, \quad 0.140]$$

$$\mathbf{D}_c = \mathrm{diag}[0.170, \quad 0.265, \quad 0.565]$$

$$\mathbf{D}_r^{-1}\mathbf{X} = \begin{bmatrix} 0.647 & 0.059 & 0.294 \\ 0.086 & 0.292 & 0.622 \\ 0.152 & 0.264 & 0.584 \\ 0.196 & 0.304 & 0.500 \end{bmatrix}$$

$$\mathbf{D}_c^{-1}\mathbf{X}^T = \begin{bmatrix} 0.324 & 0.235 & 0.279 & 0.162 \\ 0.019 & 0.509 & 0.311 & 0.160 \\ 0.044 & 0.509 & 0.323 & 0.124 \end{bmatrix}.$$

The generalized SVD of $\mathbf{X}$ is given by

$$\mathbf{X} = \begin{bmatrix} 0.085 & 0.269 & -0.050 \\ 0.463 & -0.255 & -0.166 \\ 0.313 & -0.036 & -0.131 \\ 0.140 & 0.021 & 0.346 \end{bmatrix} \begin{bmatrix} 1.0 & 0 & 0 \\ 0 & 0.403 & 0 \\ 0 & 0 & 0.047 \end{bmatrix}$$

$$\times \begin{bmatrix} 0.170 & 0.265 & 0.565 \\ 0.374 & -0.153 & -0.222 \\ 0.029 & 0.414 & -0.443 \end{bmatrix}$$

$$\mathbf{U} = \begin{bmatrix} 1 & 3.167 & -0.591 \\ 1 & -0.550 & -0.358 \\ 1 & -0.116 & -0.418 \\ 1 & 0.153 & 2.474 \end{bmatrix}$$

$$\mathbf{V} = \begin{bmatrix} 1 & 2.203 & 0.172 \\ 1 & -0.576 & 1.563 \\ 1 & -0.393 & -0.785 \end{bmatrix}$$

$$\mathbf{UD}_\lambda = \begin{bmatrix} 1 & 1.276 & -0.023 \\ 1 & -0.222 & -0.017 \\ 1 & -0.047 & -0.020 \\ 1 & 0.062 & 0.116 \end{bmatrix}$$

$$\mathbf{VD}_\lambda = \begin{bmatrix} 1 & 0.888 & 0.008 \\ 1 & -0.232 & 0.073 \\ 1 & -0.158 & -0.037 \end{bmatrix}.$$

There is always a singular value of unity with associated eigenvectors $\mathbf{1}$. This is easily seen since $\mathbf{D}_r^{-1}\mathbf{X1} = \mathbf{1}$, $\mathbf{D}_c^{-1}\mathbf{X}^T\mathbf{1} = \mathbf{1}$, noting the $\mathbf{1}$ vectors have differing lengths. From Section 1.4.2 the singular values in $\mathbf{D}_\lambda$ are given by the square roots of the non-zero eigenvalues of

$$(\mathbf{D}_r^{-\frac{1}{2}}\mathbf{XD}_c^{-\frac{1}{2}})(\mathbf{D}_r^{-\frac{1}{2}}\mathbf{XD}_c^{-\frac{1}{2}})^T = \mathbf{D}_r^{-\frac{1}{2}}\mathbf{XD}_c^{-1}\mathbf{X}^T\mathbf{D}_r^{-\frac{1}{2}}. \qquad (8.3)$$

These eigenvalues are the same as those of $\mathbf{D}_r^{-1}\mathbf{XD}_c^{-1}\mathbf{X}^T$. So

$$\mathbf{D}_r^{-1}\mathbf{XD}_c^{-1}\mathbf{X}^T\mathbf{1} = \mathbf{D}_r^{-1}\mathbf{X1} = \mathbf{1}.$$

Hence unity is an eigenvalue and also a singular value and $\mathbf{1}$ will be the corresponding singular vector of $\mathbf{U}$. A similar argument also shows that $\mathbf{1}$ is the corresponding singular vector for $\mathbf{V}$.

The singular value of unity and its associated singular vectors $\mathbf{1}$ give rise to the so called trivial dimension and can be omitted from calculations by removal from row and column profile matrices. Thus the matrices submitted to correspondence analysis are $\mathbf{D}_r^{-1}\mathbf{X} - \mathbf{1c}^T$ and $\mathbf{D}_c^{-1}\mathbf{X}^T - \mathbf{1r}^T$, where $\mathbf{r}$ and $\mathbf{c}$ are vectors of row and column sums.

Ignoring the trivial dimension, Figure 8.2 uses the singular vectors of $\mathbf{UD}_\lambda$ to plot points representing the histological type of tumour, and the singular vectors of $\mathbf{VD}_\lambda$ for points representing the site of the tumours. One dimensional spaces for type of tumour and site of tumour can easily be gleaned from the figure by simply ignoring the second axis. Since the first singular value is nearly nine times as large as the second, one dimensional spaces adequately represent the types and sites of tumour.

The figure shows that Hutchinson's melanotic freckle stands well away from the other types of tumour, and the head and neck away from the other two sites. The row profile matrix confirms that this should be so with 65% of Hutchinson's melanotic freckle occurring on the head and neck, while the other three tumour types each

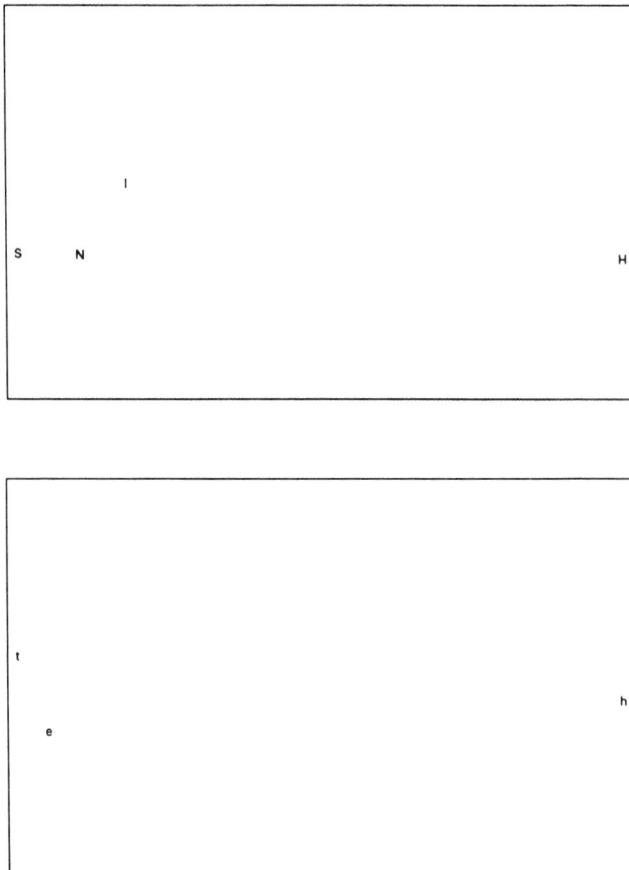

Figure 8.2 *Correspondence analysis of the cancer data. First space – histological type, second space – site of tumour.*

have over 50% of their occurrences at the extremities. The column profiles show that the head and neck is a common site for all four types of tumour, while the trunk and the extremities rarely have Hutchinson's melanotic freckle and with 50% of their occurrences being superficial spreading melanoma.

These results can be compared with those using the SVD of

X. For tumour type the same ordering in one dimension is obtained, although the direction of the axis has been reversed nonconsequentially. However the results for the site of tumour are very different for those from the SVD analysis.

8.3.2 Inertia

A measure of the dispersion of the row profiles is given by the "total inertia", a term taken from its physical counterpart. First consider the quantity $n^{-1}X^2$, where X^2 is the usual quantity calculated from a two-way contingency table, under the assumption of independent rows and columns,

$$n^{-1}X^2 = n^{-1}\sum_i\sum_j \frac{\left(x_{ij} - \frac{x_{i+}x_{+j}}{n}\right)^2}{\frac{x_{i+}x_{+j}}{n}},$$

where x_{i+} and x_{+j} are row and column sums.

This can be written

$$n^{-1}X^2 = \sum_i\left\{x_{i+}\sum_j \frac{\left(\frac{x_{ij}}{x_{i+}} - \frac{x_{+j}}{n}\right)^2}{x_{+j}}\right\},$$

$$= \sum_i r_i(\mathbf{r}_i - \mathbf{c})^T\mathbf{D}_c^{-1}(\mathbf{r}_i - \mathbf{c}) = I,$$

where r_i is the ith row sum, $\mathbf{r}_i$ is the ith row profile in the form of a vector, $\mathbf{c}$ is the "average row profile" formed as the profile of the column sums. This is the total inertia for the row profiles. It is a weighted sum of the squared weighted distances between row profiles and the "average row profile".

Interchanging rows and columns gives the total inertia for the column profiles as

$$\sum_j c_j(\mathbf{c}_j - \mathbf{r})^T\mathbf{D}_r^{-1}(\mathbf{c}_j - \mathbf{r}),$$

and by symmetry of X^2 is equal to the total inertia for the row profiles.

Now I can be written as

$$I = \mathrm{tr}(\mathbf{D}_r(\mathbf{D}_r^{-1}\mathbf{x} - \mathbf{1}\mathbf{c}^T)\mathbf{D}_c^{-1}(\mathbf{D}_r^{-1}\mathbf{X} - \mathbf{1}\mathbf{c}^T)^T),$$

where $\mathbf{D}_r^{-1}\mathbf{X} - \mathbf{1}\mathbf{c}^T$ is the matrix of row profiles with the trivial dimension removed. Replace $\mathbf{D}_r^{-1}\mathbf{X} - \mathbf{1}\mathbf{c}^T$ by $\mathbf{D}_r^{-1}\mathbf{X}$ assuming this trivial dimension has been removed, then

$$
\begin{aligned}
I &= \operatorname{tr}(\mathbf{D}_r(\mathbf{D}_r^{-1}\mathbf{X})\mathbf{D}_c^{-1}(\mathbf{D}_r^{-1}\mathbf{X})^T) \\
&= \operatorname{tr}((\mathbf{A}\mathbf{D}_\lambda\mathbf{B}^T)\mathbf{D}_c^{-1}(\mathbf{B}\mathbf{D}_\lambda\mathbf{A}^T)\mathbf{D}_r^{-1}) \\
&= \operatorname{tr}(\mathbf{A}\mathbf{D}_\lambda^2\mathbf{A}^T\mathbf{D}_r^{-1}) = \operatorname{tr}(\mathbf{D}_\lambda^2\mathbf{A}^T\mathbf{D}_r^{-1}\mathbf{A}) \\
&= \operatorname{tr}(\mathbf{D}_\lambda^2).
\end{aligned}
$$

Hence the total inertia is equal to the sum of the squared singular values. The required dimension of the row and column profile spaces can be judged by the contribution to the total inertia by the various dimensions. Thus if k dimensional spaces are chosen the contribution to total inertia is

$$
\sum_1^k \lambda_i^2 / \sum_1^n \lambda_i^2,
$$

where n is the total number of non-unit singular values. For the cancer example the total inertia was 0.1645 and the first dimension contributed 98.6% of this.

8.4 Reciprocal averaging

Reciprocal averaging, like dual scaling, is essentially the same as correspondence analysis, although Greenacre (1984) maintains that there are differences, especially in the geometric framework of the various models. The term reciprocal averaging was first used by Hill (1973, 1974) and has since become very popular with plant ecologists. It is within this area that the theory can be well illustrated.

Suppose n different species of plants are investigated at p different sites, and to fix ideas suppose the sites are chosen for their varying exposure to extreme weather conditions, while the species of plant are chosen for their various levels of hardiness. Let $\mathbf{X} = [x_{ij}]$, where x_{ij} is the response of species i at site j. For example the ecologist may simply be interested in presence/absence ($x_{ij} = 1/0$) of the ith species at the jth site.

Let u_i be a hardiness score for the ith species. Let v_j be an exposure score for the jth site. It is assumed that the exposure score at the jth site is proportional to the mean hardiness score of the species at that site. Thus

$$v_j \propto \sum_i u_i x_{ij} / \sum_i x_{ij}.$$

Correspondingly it is assumed that the hardiness score of species i is proportional to the mean exposure score of the sites occupied by that species. Thus

$$u_i \propto \sum_j v_j x_{ij} / \sum_j x_{ij}.$$

Reciprocal averaging then solves the two equations

$$\rho u_i = \sum_j v_j x_{ij} / r_i \quad (i = 1, \ldots, n) \tag{8.4}$$

$$\rho v_j = \sum_i u_i x_{ij} / c_j \quad (j = 1, \ldots, p), \tag{8.5}$$

where $r_i = \sum_i x_{ij}$, $c_j = \sum_j x_{ij}$, and ρ is a scaling parameter.

8.4.1 Algorithm for solution

A trivial solution is $\rho = 1$, $u_i = 1$, $(i = 1, \ldots, n)$, $v_j = 1$, $(j = 1, \ldots, p)$ (cf. the trivial dimension of the theory of correspondence analysis, with singular value unity, and singular vectors $\mathbf{1}$). This trivial solution is removed from the data by transforming to $x_{ij} - r_i c_j / x_{..}$, and then solving equations (8.4) and (8.5) iteratively.

Hill (1973) gives an algorithm for the solution. Choose an initial set of exposure scores placed in a vector $\mathbf{v_0}$. The scores are scaled so that the smallest is zero and the largest unity, say. Let $\rho = 1$ and calculate hardiness scores $\mathbf{u_1}$ from (8.4). Use these in (8.5) to obtain updated exposure scores $\mathbf{v_1}$ which are then scaled again to have minimum zero and maximum unity. This process is continued until convergence. The value of ρ is calculated as the factor required for the scaling of the final scores. The value of ρ and the two sets of scores give rise to a first axis. This first axis can be "subtracted" from the incidence matrix and then the whole procedure repeated to find a second axis, and so forth. However since reciprocal averaging is related to correspondence analysis the axes are more easily found as eigenvalues and eigenvectors of various matrices as discussed below.

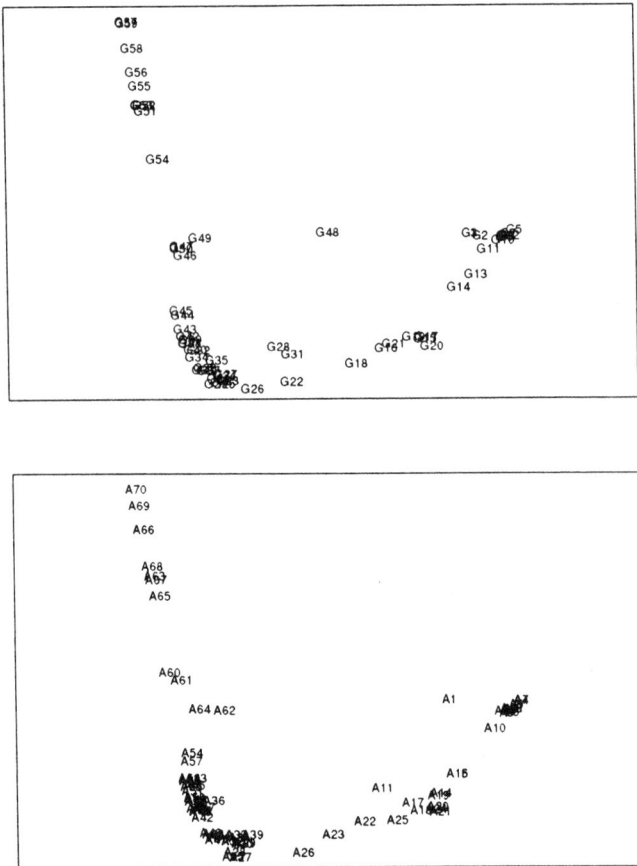

Figure 8.3 *Reciprocal averaging of the Münsingen data. Upper figure –
graves, lower figure – artefacts.*

8.4.2 An example: the Münsingen data

Hodson's Münsingen data have been a popular candidate for reciprocal averaging. The contents of various ancient graves at La Tène
cemetery at Münsingen-Rain in Switzerland were recorded. From
this an incidence matrix, $\mathbf{X}$, is formed where each row represents
a grave and each column an artefact - pottery, jewellery, etc. Then

$[\mathbf{X}]_{ij} = 1$ if the ith grave contains an example of the jth artefact, and zero otherwise. Kendall (1971) gives the data and an analysis.

Figure 8.3 shows a plot of the grave scores and the artefact scores recovered as the first two axes from reciprocal averaging. The grave scores form a "horseshoe" as do the artefacts, a phenomenon discussed by Kendall. It is possible that taking the graves and artefacts in order around their horseshoes will give an age ordering to the graves and artefacts.

8.4.3 The correspondence analysis connection

If all the dimensions found by reciprocal averaging are considered simultaneously, then the method is seen to be equivalent to correspondence analysis. Equations (8.4) and (8.5) can be written as

$$\rho\mathbf{u} = \mathbf{D}_r^{-1}\,\mathbf{X}\mathbf{v}$$
$$\rho\mathbf{v} = \mathbf{D}_c^{-1}\mathbf{X}^T\mathbf{u},$$

where $\mathbf{D}_r = \mathrm{diag}(\sum_j x_{ij})$, and $\mathbf{D}_c = \mathrm{diag}(\sum_i x_{ij})$. Then

$$\rho\mathbf{D}_r^{\frac{1}{2}}\mathbf{u} = (\mathbf{D}_r^{-\frac{1}{2}}\mathbf{X}\mathbf{D}_c^{-\frac{1}{2}})\mathbf{D}_c^{\frac{1}{2}}\mathbf{v}, \qquad (8.6)$$
$$\rho\mathbf{D}_c^{\frac{1}{2}}\mathbf{v} = (\mathbf{D}_r^{-\frac{1}{2}}\mathbf{X}\mathbf{D}_c^{-\frac{1}{2}})^T\mathbf{D}_r^{\frac{1}{2}}\mathbf{u}. \qquad (8.7)$$

Substituting equation (8.7) into equation (8.6) gives

$$\rho^2(\mathbf{D}_r^{\frac{1}{2}}\mathbf{u}) = (\mathbf{D}_r^{-\frac{1}{2}}\mathbf{X}\mathbf{D}_c^{-\frac{1}{2}})(\mathbf{D}_r^{-\frac{1}{2}}\mathbf{X}\mathbf{D}_c^{-\frac{1}{2}})^T\mathbf{D}_r^{\frac{1}{2}}\mathbf{u}.$$

Hence ρ^2 is an eigenvalue of

$$(\mathbf{D}_r^{-\frac{1}{2}}\mathbf{X}\mathbf{D}_c^{-\frac{1}{2}})(\mathbf{D}_r^{-\frac{1}{2}}\mathbf{X}\mathbf{D}_c^{-\frac{1}{2}})^T = \mathbf{D}_r^{-\frac{1}{2}}\mathbf{X}\mathbf{D}_c^{-1}\mathbf{X}^T\mathbf{D}_r^{-\frac{1}{2}},$$

and has an associated eigenvector $(\mathbf{D}_r^{\frac{1}{2}}\mathbf{u})$. But these are just the square of the singular value and the associated singular vector in (8.3). Likewise, substituting equation (8.6) into equation (8.7) gives ρ^2 as an eigenvalue of $\mathbf{D}_c^{-\frac{1}{2}}\mathbf{X}^T\mathbf{D}_r^{-1}\mathbf{X}\mathbf{D}_c^{-\frac{1}{2}}$, and associated eigenvector $(\mathbf{D}_c^{\frac{1}{2}}\mathbf{v})$. Thus reciprocal averaging finds in turn all the singular values and singular vectors associated with correspondence analysis.

8.5 Multiple correspondence analysis

Correspondence analysis is eminently suited to analysing two-way contingency tables – correspondence analysis needs all the elements of the data matrix $\mathbf{X}$ to be non-negative. Correspondence analysis can also be used on three-way or higher-way contingency tables. This is achieved by using indicator variables to convert the multi-way table into a two-way table. Suppose for a k-way table the number of categories for the ith way is c_i. An indicator variable is assigned to each category of each way of the table, giving $J = \sum_1^k c_i$ indicator variables in total. Each individual count out of the total count of n, then forms a row of an $n \times J$ table with the J indicator variables forming the columns. Each row of the new table will have k values of unity and $J - k$ of zero. An indicator variable has value unity if the individual count is in the corresponding category of the original table. For example the cancer data, although already a two-way table, can be put in this new form, giving a 400×7 table. Let the indicator variables be assigned: $I_1 = H$, $I_2 = S$, $I_3 = N$, $I_4 = I$, $I_5 = h$, $I_6 = t$, and $I_7 = e$. Then the first 22 rows of the table would be identical and equal to $(1, 0, 0, 0, 1, 0, 0)$. Then follows 2 rows of $(1, 0, 0, 0, 0, 1, 0)$, etc., the table ending with 28 rows of $(0, 0, 0, 1, 0, 0, 1)$.

Figure 8.4 shows the correspondence analysis output for the cancer data using the indicator matrix. It can be seen that the positions of the four tumour types and three sites occupy similar positions to those from their previous analysis, noting however that this time the two axes have not been scaled by their respective singular values.

The eigenvalues of the two methods, i.e. the first using the usual correspondence analysis technique, and the second making use of an indicator matrix, are related by

$$\rho = (2\rho_I - 1)^2,$$

where ρ is an eigenvalue based on the original data matrix, and ρ_I an eigenvalue based on the indicator matrix. See Greenacre (1984) for further details.

For a k-way contingency table, the indicator matrix can be written $\mathbf{Z} = [\mathbf{Z}_1, \ldots, \mathbf{Z}_k]$ where $\mathbf{Z}_i$ is an $n \times c_i$ matrix containing the c_i indicator variables for the ith way of the table. The matrix $B = \mathbf{Z}^T \mathbf{Z}$ is called the Burt matrix and contains the submatrices

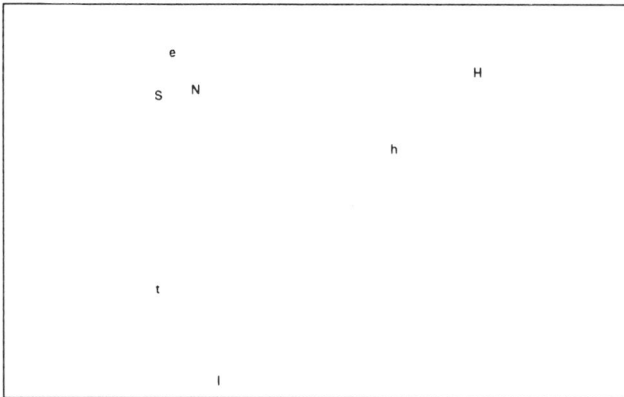

Figure 8.4 *Correspondence analysis of the cancer data represented by an indicator matrix.*

$\mathbf{Z}_i^T \mathbf{Z}_j$, the two-way contingency tables based on the ith and jth variables. Thus

$$
\mathbf{B} = \begin{bmatrix}
\mathbf{Z}_1^T \mathbf{Z}_1 & \mathbf{Z}_1^T \mathbf{Z}_2 & \cdots & \mathbf{Z}_1^T \mathbf{Z}_k \\
\mathbf{Z}_2^T \mathbf{Z}_1 & \mathbf{Z}_2^T \mathbf{Z}_2 & \cdots & \mathbf{Z}_2^T \mathbf{Z}_k \\
\vdots & \vdots & \ddots & \vdots \\
\mathbf{Z}_k^T \mathbf{Z}_1 & \mathbf{Z}_k^T \mathbf{Z}_2 & \cdots & \mathbf{Z}_k^T \mathbf{Z}_k
\end{bmatrix}.
$$

The submatrices $\mathbf{Z}_i^T \mathbf{Z}_i$ on the diagonal are simply diagonal matrices of column sums.

8.5.1 A three-way example

The three-way data in Table 8.2 are taken from Plackett (1981) and relate infant losses (e.g. stillbirths) for mothers to birth order and to whether there is a problem child in the family. A Burt matrix was found from the data which was then subjected to multiple correspondence analysis. Results are shown in Figure 8.5. Plackett's analysis indicated that only birth order affects the infant losses. This is confirmed in Figure 8.5 since the "problem/control axis" is nearly perpendicular to the "losses/none axis" with no pair of

Table 8.2 *Infant losses in relation to birth order and problem children. P – problem, C – controls.*

| Numbers of mothers with | Birth order | | | | | |
| | 2 | | 3–4 | | 5+ | |
	P	C	P	C	P	C
Losses	20	10	26	16	27	14
None	82	54	41	30	22	23

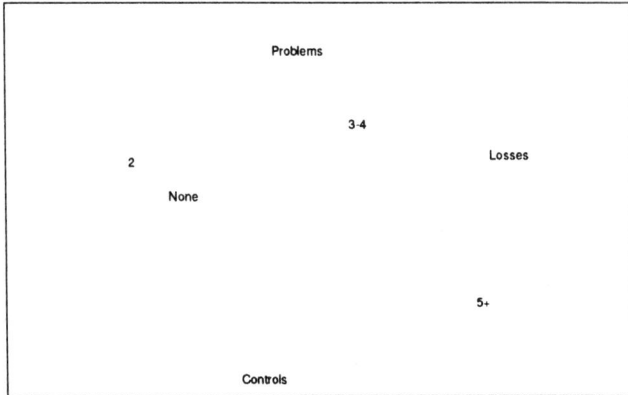

Figure 8.5 *Correspondence analysis of the three-way infant loss data.*

points being close. The "birth order axis" 2/3–4/5+ is more aligned with the losses/none axis indicating a relationship between these two, although this relationship would have appeared stronger if the "5+" had been closer to "losses".

For a comprehensive introduction to correspondence analysis, see Benzécri (1992). Some more recent articles on the subject are Tenenhaus and Young (1985), Greenacre and Hastie (1987), Choulakian (1988), Greenacre (1988), de Leeuw and van der Heijden (1988), Gower (1990). For correspondence analysis linked to loglinear models, see van der Heijden and de Leeuw (1985), van der Heijden and Worsley (1988), van der Heijden *et al.* (1989) and van der Heijden and Meijerink (1989). Bénasséni (1993) considers some perturbational aspects in correspondence analysis. Gilula and Ritov (1990), Pack and Jolliffe (1992) and Krzanowski (1993) consider some inferential aspects of correspondence analysis.

CHAPTER 9

Individual differences models

9.1 Introduction

Data analysed so far have generally been two-way, one- or two-mode data. This chapter investigates models for three-way, two-mode data, in particular for dissimilarities $\delta_{rs,i}$ where the suffices r and s refer to one set of objects and i to another. For example N judges might each be asked their opinions on n objects or stimuli, from which N separate dissimilarity matrices are derived. The (r, s)th element of the ith dissimilarity matrix would be $\delta_{rs,i}$. Another example might be the production of a dissimilarity matrix each year for schools in a certain region, based on exam results. Then r and s refer to the rth and sth schools, and i is a time index. Individual differences modelling attempts to analyse such data taking into account the two different modes. For convenience the suffix i will refer to "individuals" rather than any other objects such as points in time. The whisky tasting experiment discussed in Chapter 1 is another example with a dissimilarity matrix produced for each of the N judges.

There were two basic approaches in the early work in this area. The first was to average over individuals, the second to compare results individual by individual. For example metric or nonmetric MDS could be used on the dissimilarity matrix obtained by averaging dissimilarities over i, or alternatively by carrying out an analysis for each individual and then attempting a comparison.

9.2 The Tucker-Messick model

Tucker and Messick (1963) addressed the problems with the two early approaches to individual differences scaling, namely that averaging over individuals loses much information regarding the individual responses, and that comparing several different scalings can be a very difficult task. Tucker and Messick (see also Cliff (1968)) suggested placing the dissimilarities $\{\delta_{rs,i}\}$ into a matrix,

X, with rows given by all the $\frac{1}{2}n(n-1)$ possible stimulus-pairs and columns given by the N individuals. Essentially the singular valued decomposition (SVD) of **X** is then found,

$$\mathbf{X} = \mathbf{U}\mathbf{\Lambda}\mathbf{V}^T$$

and then the p dimensional least squares approximation to **X**,

$$\hat{\mathbf{X}}_p = \mathbf{U}_p\mathbf{\Lambda}_p\mathbf{V}_p^T.$$

The matrix $\mathbf{U}_p$ gives the principal coordinates in a space for the pairs of stimuli, the matrix $\mathbf{\Lambda}_p\mathbf{V}^T$ gives the principal coordinates in a space for the individuals.

9.3 INDSCAL

Carroll and Chang (1970) proposed a metric model comprising two spaces: a group stimulus space and a subjects (or individuals) space, both of chosen dimension p. Points in the group stimulus space represent the objects or stimuli, and form an "underlying" configuration. The individuals are represented as points in the subjects space. The coordinates of each individual are the weights required to give the weighted Euclidean distances between the points in the stimulus space, the values that best represent the corresponding dissimilarities for that individual. Hence the acronym INDSCAL – INdividual Differences SCALing.

Let the points in the group stimulus space be given by x_{rt} ($r = 1, \ldots, n; t = 1, \ldots, p$). Let the points in the individuals space have coordinates w_{it} ($i = 1, \ldots, N; t = 1, \ldots, p$). Then the weighted Euclidean distance between stimuli r and s, for the ith individual is

$$d_{rs,i} = \left\{ \sum_{t=1}^{p} w_{it}(x_{rt} - x_{st})^2 \right\}^{\frac{1}{2}}.$$

The individual weights $\{w_{it}\}$ and stimuli coordinates $\{x_{rt}\}$ are then sought that best match $\{d_{rs,i}\}$ to $\{\delta_{rs,i}\}$.

9.3.1 The algorithm for solution

As with metric scaling of Chapter 2, dissimilarities $\{\delta_{rs,i}\}$ are converted to distance estimates $\{d_{rs,i}\}$ and then $\{w_{it}\}$, $\{x_{rt}\}$ are found by least squares. The distances associated with each individual are doubly centred giving matrices $\mathbf{B}_i$, where

$$[\mathbf{B}_i]_{rs} = b_{rs,i} = \sum_{t=1}^{p} w_{it} x_{rt} x_{st}$$

$$= -\frac{1}{2}\left(d_{rs,i}^2 - \frac{1}{n}\sum_{r=1}^{N} d_{rs,i}^2 - \frac{1}{n}\sum_{s=1}^{N} d_{rs,i}^2 + \frac{1}{n^2}\sum_{r=1}^{N}\sum_{s=1}^{N} d_{rs,i}^2\right)$$

$$= \mathbf{H}\mathbf{A}_i\mathbf{H},$$

and $[\mathbf{A}_i]_{rs} = a_{rs,i} = -\frac{1}{2}d_{rs,i}^2$. Least squares estimates of $\{w_{it}\}$ and $\{x_{rt}\}$ are then found by minimizing

$$S = \sum_{r,s,i}\left(b_{rs,i} - \sum_{t=1}^{p} w_{it} x_{rt} x_{st}\right)^2. \tag{9.1}$$

Carroll and Chang's algorithm uses a recursive least squares approach. Firstly, superscripts L and R (Left and Right) are placed on x_{rt} and x_{st} respectively in equation (9.1) to distinguish two estimates of the coordinates of the points in the group stimulus space, which converge to a common estimate. Thus equation (9.1) is

$$S = \sum_{r,s,i}\left(b_{rs,i} - \sum_{t=1}^{p} w_{it} x_{rt}^L x_{st}^R\right)^2.$$

The quantity S is firstly minimized with respect to $\{w_{it}\}$ for fixed $\{x_{rt}^L\}$, $\{x_{st}^R\}$. This is easily achieved if $\{x_{rt}^L x_{st}^R\}$ forms an $n^2 \times p$ matrix $\mathbf{G}$, where $[\mathbf{G}]_{\alpha t} = x_{rt}^L x_{st}^R$, with $\alpha = n(r-1) + s$, and $\{b_{rs,i}\}$ forms the $N \times n^2$ matrix $\mathbf{F}$ where $[\mathbf{F}]_{i\alpha} = b_{rs,i}$. Let the $N \times p$ matrix $\mathbf{W}$ be given by $[\mathbf{W}]_{it} = w_{it}$. Then the least squares estimate of $\mathbf{W}$ is given by

$$\hat{\mathbf{W}} = \mathbf{F}\mathbf{G}(\mathbf{G}^T\mathbf{G})^{-1}.$$

Next a least squares estimate of $\{x_{rt}^L\}$ is found for fixed $\{w_{it}\}$, $\{x_{st}^R\}$. Let $\mathbf{G}$ now be the $Nn \times p$ matrix $[\mathbf{G}]_{\alpha t} = w_{it} x_{st}^R$, where now $\alpha = n(i-1) + s$. Let $\mathbf{F}$ be the $n \times Nn$ matrix $[\mathbf{F}]_{\alpha\beta} = b_{rs,i}$ where $\alpha = r$, $\beta = n(i-1) + s$. Let $\mathbf{X}^L$ be the $n \times p$ matrix $[\mathbf{X}^L]_{rt} = x_{rt}^L$. Then the least squares estimate of $\mathbf{X}^L$ for fixed $\{w_{it}\}$, $\{x_{st}^R\}$ is

$$\hat{\mathbf{X}}^L = \mathbf{F}\mathbf{G}(\mathbf{G}^T\mathbf{G})^{-1}.$$

This last step is now repeated interchanging $\{x_{rt}^L\}$ and $\{x_{st}^R\}$ to find the least squares estimate $\mathbf{X}^R$ of $\{x_{st}^R\}$ for fixed $\{w_{it}\}$, $\{x_{rt}^L\}$.

The process is repeated until convergence of $\hat{\mathbf{X}}^L$ and $\hat{\mathbf{X}}^R$. Carroll and Chang point out that $\hat{\mathbf{X}}^L$, $\hat{\mathbf{X}}^R$ converge only up to a diagonal transformation,

$$\mathbf{X}^L = \mathbf{X}^R \mathbf{C}$$

where $\mathbf{C}$ is a $p \times p$ diagonal matrix of non-zero entries. This is because $\sum_{t=1}^{p} w_{it} x_{rt} x_{st}$ can be replaced by $\sum_{t=1}^{p} (w_{it}/c_t) x_{rt}^L (x_{st} c_t)$ in equation (9.1), and hence the minimum sum of squares is not affected by $\{c_t\}$. To overcome this, the final step in the procedure is to set $\hat{\mathbf{X}}^L$ equal to $\hat{\mathbf{X}}^R$ and compute $\hat{\mathbf{W}}$ for a last time.

Notice one property of the INDSCAL model, that the dimensions of the resulting spaces are unique. Configurations cannot be translated or rotated. This implies that the dimensions may possibly be interpreted.

Normalization
Carroll and Chang address two normalization questions. The first is the weighting of the contributions to the analysis by the different individuals. Unless there are specific reasons to do so, they suggest that individuals are weighted equally, which is achieved by normalizing each individual's sum of squared scalar products, $\sum_{r,s} b_{rs,i}^2$. Secondly the final solution for the stimulus space needs to be normalized since S in equation (9.1) is invariant to dilation of the configuration of points in the stimulus space with a corresponding shrinking in the subject space. Normalization can be carried out by setting the variance of the projections of the points on each axis equal to unity.

9.3.2 Identifying groundwater populations

One of the data sets (number 17) in Andrews and Hertzberg (1985), concerns the estimation of the uranium reserves in the United States of America. These data will be subjected to analysis by INDSCAL. The data consist of twelve measurements made on groundwater samples taken at various sites. The variables are:

uranium (U); arsenic (AS); boron (B); barium (BA); molybdenum (MO); selenium (SE); vanadium (V); sulphate (SO4); total alkalinity (T_AK); bircarbonate (BC); conductivity (CT) and pH (PH).

(i)

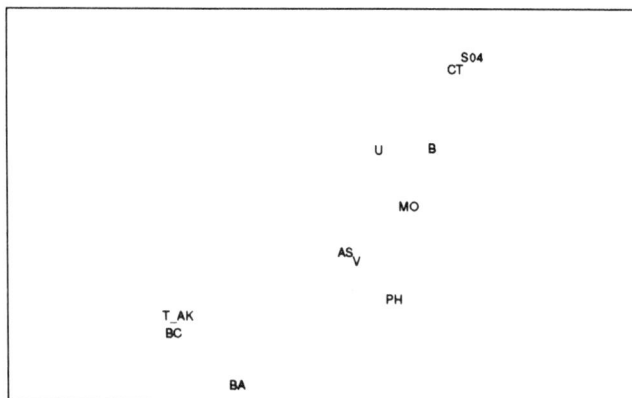

(ii)

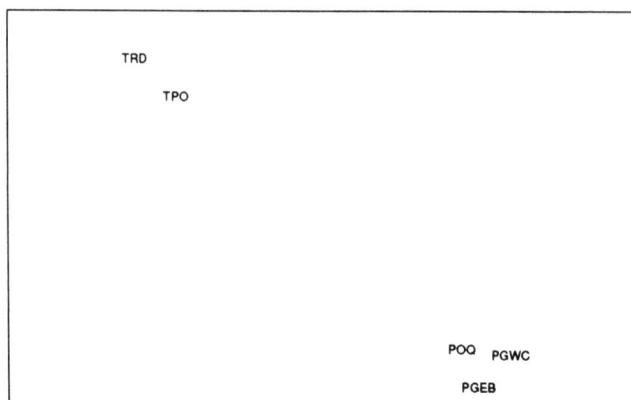

Figure 9.1 *INDSCAL analysis of groundwater samples, (i) group stimulus space, (ii) subject space.*

Each groundwater sample was initially classified as coming from one of five rock formations:

Orgallala Formation (TPO); Quartermaster Group (POQ); Whitehorse and Cloud Chief Group (PGWC); El Reno Group and Blaire Formation (PGEB); and Dockum Formation (TRD).

For each of the five classes, the sample correlation matrix was used to give dissimilarities between the variables, using the transformation $\delta_{rs} = (1 - \rho_{rs})^{\frac{1}{2}}$. One variable (SE) was left out of the analysis since for most samples it was barely measurable. The five sets of dissimilarities $\{\delta_{rs,i}\}$ were then subjected to analysis by INDSCAL using a two dimensional group stimulus space and subject space.

Figure 9.1(i) shows the group stimulus space for the eleven remaining variables. Interesting groupings are {sulphate (SO4) and conductivity (CT)}, {total alkalinity (T_AK), bicarbonate (BC), and barium (BA) }, and {arsenic (AS), vanadium (V), uranium (U), boron (B), molybdenum (MO), and pH (PH)}.

Figure 9.1(ii) shows the subject space, where two groups can be clearly seen, {Ogallala Formation (TPO) and Dockum Formation (TRD)} and {Quartermaster Group (POQ), Whitehorse and Cloud Chief Group (PGWC) and El Reno Group and Blaire Formation (PGEB)}. The first group tends to shrink the group stimulus space along the first dimension and stretch it along the second. The second group does the opposite.

9.3.3 Extended INDSCAL models

MacCallum (1976) carried out a Monte Carlo investigation of INDSCAL where for the ith individual, the angle between the axes in the group stimulus space was changed from $90°$ to θ_i, introducing error. He concluded that INDSCAL was susceptible to the assumption that individuals perceive the dimensions of the group stimulus space to be orthogonal. IDIOSCAL, a generalization of INDSCAL, can overcome this problem.

Winsberg and Carroll (1989a,b) extend the INDSCAL model to

$$d_{rs,i} = \left\{ \sum_{t=1}^{p} w_{it}(x_{rt} - x_{st})^2 + u_i(s_r - s_s) \right\}^{\frac{1}{2}},$$

where s_r is the "specificity" of the rth stimulus and u_i is the propensity of the ith individual towards specificities. The specificity for a stimulus can be thought of as a dimension solely for that stimulus. They use a maximum likelihood approach to fit the model.

Winsberg and De Soete (1993) adapt INDSCAL and assume the N individuals each belongs to a latent class or subpopulation. The probability that an individual belongs to latent class l is p_l

$(1 \leq l \leq L)$. For those individuals in latent class l, their dissimilarities $\{\delta_{rs,i}\}$ are assumed to follow a common multivariate normal distribution. The coordinates of the points in the group stimulus space and the weights in the subject space are then found by maximum likelihood. They call their model CLASCAL.

9.4 IDIOSCAL

Carroll and Chang (1972) generalized their INDSCAL model to the IDIOSCAL model (Individual DIfferences in Orientation SCALing). They used the weighted Euclidean distance between stimuli r and s, for individual i

$$d_{rs,i} = \left\{ \sum_{t=1}^{p} \sum_{t'=1}^{p} (x_{rt} - x_{st}) w_{tt',i} (x_{rt'} - x_{st'}) \right\}^{\frac{1}{2}}.$$

Here $\mathbf{W}_i$ is a symmetric positive definite or semi-definite matrix of weights, $[\mathbf{W}_i]_{tt'} = w_{tt',i}$.

It is easily seen that

$$b_{rs,i} = \sum_{t} \sum_{t'} x_{rt} w_{tt',i} x_{st'},$$

and

$$\mathbf{B}_i = \mathbf{X} \mathbf{W}_i \mathbf{X}^T.$$

The IDIOSCAL model thus allows the group stimulus space to be manipulated to a further degree by individuals than the INDSCAL model, with various rotations and dilations of axes being allowed. Carroll and Wish (1974) give a good account of models which arise from IDIOSCAL using a suitable choice of $\mathbf{W}_i$. A summary of these models is given.

INDSCAL
When $\mathbf{W}_i$ is restricted to being a diagonal matrix, IDIOSCAL reduces to INDSCAL.

Carroll-Chang decomposition of $\mathbf{W}_i$
The spectral decomposition of $\mathbf{W}_i$ is used as an aid to interpretation. Thus

$$\mathbf{W}_i = \mathbf{U}_i \boldsymbol{\Lambda}_i \mathbf{U}_i^T,$$

where $U_i U_i^T = I$, $\Lambda = \mathrm{diag}(\lambda_{ij})$, $(j = 1, \ldots, p)$, and

$$B_i = X U_i \Lambda_i U_i^T X^T$$
$$= (X U_i \Lambda_i^{\frac{1}{2}})(X U_i \Lambda_i^{\frac{1}{2}})^T,$$

which gives the interpretation of the ith individual's configuration as an orthogonal rotation, U_i, of the group stimulus space, followed by a rescaling of the axes by $\lambda_{ij}^{\frac{1}{2}}$. Unfortunately the orthogonal transformation is not unique, since for any orthogonal matrix V,

$$(U_i \Lambda_i^{\frac{1}{2}} V)(U_i \Lambda_i^{\frac{1}{2}} V)^T = (U_i \Lambda_i U_i) = W_i.$$

Tucker-Harshman decomposition of W_i
Tucker (1972) and Harshman (1972) suggested the decomposition

$$W_i = D_i R_i D_i,$$

where D_i is a diagonal matrix, and R_i is a symmetric matrix with diagonal elements all equal to unity. The matrix R_i can be interpreted as a "correlation" matrix and D_i as a diagonal matrix of "standard deviations". If $R_i = R$ $(i = 1, \ldots, N)$ then the model reduces to Harshman's PARAFAC-2 model.

9.5 PINDIS

The PINDIS model (Procrustean INdividual DIfferences Scaling) was developed along the lines of the older methods of individual differences scaling where scaling for each individual is carried out separately and then an overall comparison made. The model was developed after INDSCAL but has not been so popular. Relevant references are Borg (1977), Lingoes and Borg (1976, 1977, 1978).

PINDIS asumes that a scaling has been carried out for each individual by some method producing a configuration matrix X_i in each case. The actual scaling method is immaterial as far as PINDIS is concerned. The configurations X_i are then compared using Procrustes analysis. First a centroid configuration, Z, is established in a similar manner to that suggested by Gower (1975) (see Chapter 5) and then a hierarchy of translation, rotation, dilation models are applied to the configurations X_i, to transform them individually as best as possible to the centroid configuration. The Procrustes statistic is used as an indication of the appropriate type of translation, rotation, dilation. The centroid configuration then represents the group stimulus space and the rotations etc. for the individuals represent the subjects space.

Firstly all the N configurations $\mathbf{X}_i$ are centred at the origin and then dilated to have mean squared distance to the origin equal to unity, i.e. $\mathrm{tr}(\mathbf{X}_i^T \mathbf{X}_i) = 1$. The $\mathbf{X}_2$ configuration is rotated to the $\mathbf{X}_1$ configuration giving the first estimate of $\mathbf{Z}$ as $\frac{1}{2}(\mathbf{X}_1 + \mathbf{X}_2)$. Next $\mathbf{X}_3$ is rotated to $\mathbf{Z}$ and then a weighted average of $\mathbf{Z}$ and $\mathbf{X}_3$ gives the next estimate of $\mathbf{Z}$. This process is repeated until all the $\mathbf{X}_i$ configurations have been used.

Next the N configurations are each rotated to $\mathbf{Z}$ and a goodness of fit index calculated as

$$h = \frac{1}{N} \sum_i (1 - R_i^2(\mathbf{X}_i, \mathbf{Z}))^{\frac{1}{2}},$$

where $R(\mathbf{X}_i, \mathbf{Z})^2$ is the Procrustes statistic when $\mathbf{X}_i$ is rotated to $\mathbf{Z}$.

The average of the newly rotated $\mathbf{X}_i$ configurations gives the next updated esimate of the centroid $\mathbf{Z}$, and the goodness of fit index is recalculated. This procedure is repeated until h converges. The resulting $\mathbf{Z}$ is the centroid configuration.

The procedure so far has given the basic model. The centroid configuration $\mathbf{Z}$ is the group stimulus space, and the Procrustes rigid rotations $\mathbf{R}_i$ needed to rotate the individual configurations $\mathbf{X}_i$ to the centroid $\mathbf{Z}$ form the subject space. The rigididity of the rotations is now relaxed and various models tried. The hierarchy of models is as follows, starting with the basic model.

1. Basic model: Rigid rotations only. The quantity

$$R_1(\mathbf{X}_i, \mathbf{Z}) = \sum_i \mathrm{tr}(\mathbf{X}_i \mathbf{R}_i - \mathbf{Z})^T (\mathbf{X}_i \mathbf{R}_i - \mathbf{Z})$$

is minimized for each individual configuration $\mathbf{X}_i$.

2. Dimension weighting: The dimensions of the group stimulus space are weighted. The quantity to be minimized is

$$R_2(\mathbf{X}_i, \mathbf{Z}) = \sum_i \mathrm{tr}(\mathbf{X}_i \mathbf{R}_i - \mathbf{Z}\mathbf{S}\mathbf{W})^T (\mathbf{X}_i \mathbf{R}_i - \mathbf{Z}\mathbf{S}\mathbf{W})$$

where $\mathbf{S}^T \mathbf{S} = \mathbf{I}$, and $\mathbf{W}$ is a diagonal matrix. Here the centroid configuration $\mathbf{Z}$ is allowed to be rotated by $\mathbf{S}$ before weights are applied to the axes.

3. Idiosyncratic dimension weighting: The weighting of dimensions of the group stimulus space can be different for each individual. The quantity

$$R_3(\mathbf{X}_i, \mathbf{Z}) = \sum_i \text{tr}(\mathbf{X}_i\mathbf{R}_i - \mathbf{ZSW}_i)^T(\mathbf{X}_i\mathbf{R}_i - \mathbf{ZSW}_i)$$

is minimized for each individual configuration $\mathbf{X}_i$.

4. Vector weighting: Each stimulus in the group stimulus space is allowed to be moved along the line through the origin to the stimulus before rotation occurs. The quantity to be minimized is

$$R_4(\mathbf{X}_i, \mathbf{Z}) = \sum_i \text{tr}(\mathbf{X}_i\mathbf{R}_i - \mathbf{V}_i\mathbf{Z})^T(\mathbf{X}_i\mathbf{R}_i - \mathbf{V}_i\mathbf{Z}).$$

5. Vector weighting, individual origins: This is the same as model 4 except that the origin of $\mathbf{Z}$ for each individual can be moved to an advantageous position. The quantity to be minimized is

$$R_5(\mathbf{X}_i, \mathbf{Z}) = \sum_i \text{tr}(\mathbf{X}_i\mathbf{R}_i - \mathbf{V}_i(\mathbf{Z} - \mathbf{1}t_i^T))^T(\mathbf{X}_i\mathbf{R}_i - \mathbf{V}_i(\mathbf{Z} - \mathbf{1}t_i^T))$$

where $\mathbf{t}_i$ is the translation vector for the centroid for the ith individual.

6. Double weighting: This allows both dimensional and vector weighting. The quantity

$$R_6(\mathbf{X}_i, \mathbf{Z}) = \sum_i \text{tr}(\mathbf{X}_i\mathbf{R}_i - \mathbf{V}_i(\mathbf{Z} - \mathbf{1}t_i^T)\mathbf{W}_i)^T$$

$$\times (\mathbf{X}_i\mathbf{R}_i - \mathbf{V}_i(\mathbf{Z} - \mathbf{1}t_i^T)\mathbf{W}_i)$$

is minimized for each individual configuration $\mathbf{X}_i$.

The models form a hierarchy with the first model always fitting the worst and the last model the best. Choice of model is made by assessing the improvement in fit made by going from one model to another in the hierarchy. Langeheine (1982) evaluated the measures of fit for the various models.

CHAPTER 10

ALSCAL and SMACOF

In this chapter two significant developments in multidimensional scaling are discussed, ALSCAL and SMACOF. Both are alternatives to the gradient methods of the minimization of stress.

10.1 ALSCAL

Takane, Young and de Leeuw (1977) developed ALSCAL (Alternating Least squares SCALing) along with other uses of the alternating least squares technique (see Young, de Leeuw and Takane (1976), and de Leeuw, Young and Takane (1976)). The attraction of ALSCAL is that it can analyse data that are: (i) nominal, ordinal, interval, or ratio; (ii) complete or have missing observations; (iii) symmetric or asymmetric; (iv) conditional or unconditional; (v) replicated or unreplicated; (vi) continuous or discrete – a Pandora's box!

An outline to the theory of ALSCAL is given, following Takane *et al.* (1977).

10.1.1 The theory

As for INDSCAL of Chapter 9, assume dissimilarity data $\{\delta_{rs,i}\}$ which can be any of the types (i)-(vi) above. The scaling problem can be stated as the search for a mapping ϕ, of the dissimilarities $\{\delta_{rs,i}\}$, giving rise to a set of disparities $\{\hat{d}_{rs,i}\}$,

$$\phi[\delta_{rs,i}^2] = \hat{d}_{rs,i}^2,$$

where $\{\hat{d}_{rs,i}^2\}$ are least squares estimates of $\{d_{rs,i}^2\}$ obtained by minimizing the loss function called SSTRESS and denoted by SS, where

$$SS = \sum_r \sum_s \sum_i (d_{rs,i}^2 - \hat{d}_{rs,i}^2)^2. \tag{10.1}$$

Note that SSTRESS differs from STRESS in that it uses squared distances and disparities. This is done for algorithmic convenience.

The mapping ϕ has to take into account the restrictions that occur in the particular model and type of data. There are three types of restriction: process restrictions, level restrictions and conditionality restrictions.

Process restrictions

One process restriction is used for discrete data, another for continuous data. For discrete data, observations within a particular category should be represented by the same real number under the mapping ϕ. Following Takane *et al.*, let $\sim$ represent membership of the same category. So for discrete data

$$\phi: \quad \delta_{rs,i} \sim \delta_{r's',i'} \quad \Rightarrow \quad \hat{d}_{rs,i} = \hat{d}_{r's',i'}.$$

Continuous data have to be discretized so as to make the data categorical: for example an observation of 3.7 could be considered to be in the category of all those values in the interval $[3.65, 3.75)$. The continuous restriction is then represented by

$$\phi: \quad \delta_{rs,i} \sim \delta_{r's',i'} \quad \Rightarrow \quad l \leq \hat{d}_{rs,i}, \hat{d}_{r's',i'} \leq u,$$

where $[l, u)$ is a real interval.

Level constraints

Different constraints on ϕ are needed for the type of data being analysed. For nominal data no constraint is necessary once the process restraint has been taken into consideration. For ordinal data the obvious constraint on ϕ is

$$\phi: \delta_{rs,i} \prec \delta_{r's',i} \quad \Rightarrow \hat{d}_{rs,i} \leq \hat{d}_{r's',i'}.$$

For quantitative data, $\hat{d}_{rs,i}$ is linearly related to $\delta_{rs,i}$, so that

$$\phi: \quad \hat{d}_{rs,i} = a_0 + a_1 \delta_{rs,i},$$

with $a_0 = 0$ for ratio data. Linearity can possibly be replaced by a polynomial relationship.

Conditionality constraints

Different experimental situations give rise to different conditionalities on the dissimilarities. If measurements made by different individuals are all comparable giving the unconditional case, then no constraints on ϕ are needed. If observations by different individuals are not comparable then matrix conditionality is imposed where all dissimilarities within the matrix of dissimilarities for an individual are comparable, but not between matrices. This implies that ϕ is composed of N mappings $\{\phi_i\}$, one for each individual. Similarly row conditionality gives rise to mappings $\{\phi_{ri}\}$. Here dissimilarities along a row of a matrix are comparable but not between rows. For example N judges may score the taste of p different whiskies.

10.1.2 Minimizing SSTRESS

SSTRESS in (10.1) is minimized using an alternating least squares algorithm. Each iteration of the algorithm has two phases: an optimal scaling phase and a model estimation phase. Writing SSTRESS as $SS(\mathbf{X}, \mathbf{W}, \hat{D})$, where $\mathbf{X}$ is the matrix of coordinates, $\mathbf{W}$ is the matrix of weights, and $\hat{D}$ represents the disparities $\{\hat{d}_{rs,i}\}$, then the optimal scaling phase finds the least squares disparities $\hat{D}$ for fixed $\mathbf{X}$ and $\mathbf{W}$, which is followed by the model estimation phase which calculates new coordinates $\mathbf{X}$ and weights $\mathbf{W}$ for fixed $\hat{D}$.

The optimal scaling phase
Firstly the distances $\{d_{rs,i}\}$ are calculated from current coordinate and weight matrices $\mathbf{X}, \mathbf{W}$. Then disparities $\{\hat{d}_{rs,i}\}$ are calculated. Conveniently, if all the disparities are placed in a vector $\hat{\mathbf{d}}$, and similarly the distances placed in vector $\mathbf{d}$, then

$$\hat{\mathbf{d}} = \mathbf{E}\mathbf{d},$$

where $\mathbf{E} = \mathbf{Z}(\mathbf{Z}^T\mathbf{Z})^{-1}\mathbf{Z}^T$, with $\mathbf{Z}$ depending on the type of transformation ϕ.

For ratio and interval data $\mathbf{Z}$ is a vector of squared dissimilarities $\{d_{rs,i}^2\}$. This can easily be seen by replacing $\hat{d}_{rs,i}^2$ in the SSTRESS equation (10.1) by $a + bd\delta_{ijk}^2$ and finding the least squares estimates of a and b.

For ordinal and nominal level data $\mathbf{Z}$ is a matrix of dummy variables indicating which distances must be tied to satisfy the measurement conditions. For example with dissimilarities and distances given by

$$\delta_1^2 = 1.2 \quad \delta_2^2 = 1.7 \quad \delta_3^2 = 2.4 \quad \delta_4^2 = 3.2 \quad \delta_5^2 = 3.6$$

$$d_1^2 = 3.8 \quad d_2^2 = 4.6 \quad d_3^2 = 4.2 \quad d_4^2 = 5.4 \quad d_5^2 = 5.0,$$

for the ordinal transformation, least squares monotone regression will have $\hat{d}_2^2 = \hat{d}_3^2$, $\hat{d}_4^2 = \hat{d}_5^2$, giving

$$
\mathbf{Z} = \begin{bmatrix} 1 & 0 & 0 \\ 0 & 1 & 0 \\ 0 & 1 & 0 \\ 0 & 0 & 1 \\ 0 & 0 & 1 \end{bmatrix}
$$

and

$$
\mathbf{Z}(\mathbf{Z}^T\mathbf{Z})^{-1}\mathbf{Z}^T = \begin{bmatrix} 1 & 0 & 0 & 0 & 0 \\ 0 & \frac{1}{2} & \frac{1}{2} & 0 & 0 \\ 0 & \frac{1}{2} & \frac{1}{2} & 0 & 0 \\ 0 & 0 & 0 & \frac{1}{2} & \frac{1}{2} \\ 0 & 0 & 0 & \frac{1}{2} & \frac{1}{2} \end{bmatrix}.
$$

SSTRESS can now be written as

$$
SS = \mathbf{d}^T(\mathbf{I} - \mathbf{E})\mathbf{d}
$$

and normalized SSTRESS as

$$
SS = \mathbf{d}^T(\mathbf{I} - \mathbf{E})\mathbf{d}/\mathbf{d}^T\mathbf{d}.
$$

The last step in the optimal scaling phase is to normalize the solution, firstly with respect to the configuration and weights and other parameters, and secondly with regard to SSTRESS.

Model estimation phase

The model estimation phase finds the least squares estimates of the weight matrix, $\mathbf{W}$, for the current disparity values $\{\hat{d}_{rs,i}\}$ and coordinates $\mathbf{X}$ of the points in the group stimulus space. Then the least squares estimates of $\mathbf{X}$ are found for the current disparity values and weights $\mathbf{W}$.

For the first minimization let the $\frac{1}{2}n(n-1)$ quantities $(x_{rt}-x_{st})^2$ make up the tth column $(t = 1, \ldots, p)$ of a matrix $\mathbf{Y}$. A similar $\frac{1}{2}n(n-1) \times p$ matrix $\mathbf{D}^\star$ is composed of the disparities $\{\delta_{rs,i}^2\}$. Then SSTRESS can be written

$$
SS = \mathrm{tr}(\mathbf{D}^\star - \mathbf{W}\mathbf{Y}^T)^T(\mathbf{D}^\star - \mathbf{W}\mathbf{Y}^T)
$$

and hence

$$
\mathbf{W} = \mathbf{D}^\star\mathbf{Y}(\mathbf{Y}^T\mathbf{Y})^{-1}.
$$

There can be a problem with negative estimated weights. Takane *et al.* show how these can be appropriately adjusted.

For the second minimization the SSTRESS in (10.1) now has to be minimized with respect to the coordinates **X**. Setting partial derivatives equal to zero gives rise to a series of cubic equations which can be solved using Newton-Raphson, possibly modified. The reader is referred to Takane et al. for further details. In summary the ALSCAL algorithm is as follows:

1. Find an initial configuration **X** and weights **W**.
2. Optimal scaling phase: calculate D, $D^\star$ and normalize.
3. Terminate if SSTRESS has converged.
4. Model estimation phase: minimize $SS(\mathbf{W}|\mathbf{X}, D^\star)$ over **W**; then minimize $SS(\mathbf{X}|\mathbf{W}, D^\star)$ over **X**.
5. Go to 2.

Details of further points relating to ALSCAL and reports on some Monte Carlo testing of the technique can be found in Mac-Callum (1977a, 1977b, 1978), MacCallum and Cornelius III (1977), Young and Null (1978), Young et al. (1978), Verhelst (1981) and ten Berge (1983).

To reiterate, the attraction of ALSCAL is that it is very versatile and can perform metric scaling, nonmetric scaling, multidimensional unfolding, individual differences scaling and other techniques. ALSCAL is available in the statistical computer packages SAS and SPSS. Chapter 12 discusses briefly the implementation in SPSS.

10.2 SMACOF

As an alternative to the alternating least squares method for minimizing SSTRESS, a method based on the majorization algorithm was initially proposed by de Leeuw (1977b). The method was then further refined and explored by de Leeuw and Heiser (1977, 1980), de Leeuw (1988), Heiser (1991), de Leeuw (1992) and Groenen (1993), and now has the acronym SMACOF which stands for Scaling by MAjorizing a COmplicated Function. Before describing SMACOF the majorizing algorithm is briefly described. The following relies heavily on Groenen (1993).

10.2.1 The majorization algorithm

The majorization algorithm attempts to minimize a complicated function, $f(x)$, by use of a more manageable auxiliary function

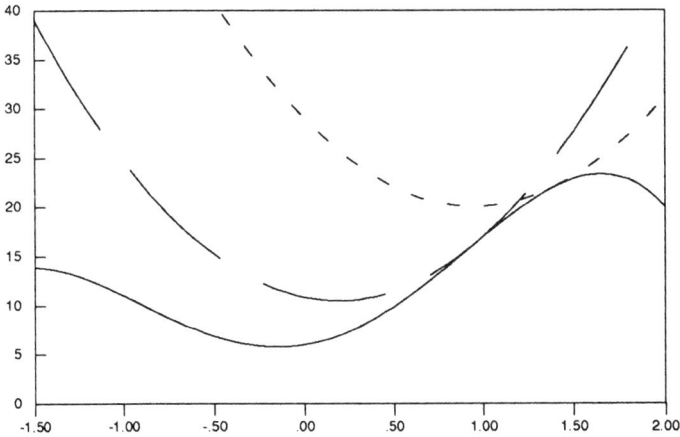

Figure 10.1 *Minimizing the function* $f(x) = 6 + 3x + 10x^2 - 2x^4$ *using the majorizing function* $g(x, y) = 6 + 3x + 10x^2 - 8xy^3 + 6y^4$. *Solid line,* f; *short dashed line,* $g(x, 1.4)$; *long dashed line,* $g(x, 0.948)$.

$g(x, y)$. The auxiliary function has to be chosen such that for each x in the domain of f

$$f(x) \leq g(x, y),$$

for a particular y in the domain of g, and also so that

$$f(y) = g(y, y).$$

So for graphs of f and g, the function g is always above the function f, and g touches f at the point $x = y$. The function g is then a majorizing function of f. This leads to an iterative scheme to minimize f. First an initial value x_0 is used to start the minimization. This then defines the appropriate majorizing function $g(x, x_0)$. This is minimized with its minimum at x_1 say. This value of x then defines the majorizing function $g(x, x_1)$. This in turn is minimized with minimum at x_2. The process is repeated until convergence.

An example

As an example of the majorizing algorithm consider minimizing the function, f, where

$$f : [-1.5, 2.0] \longrightarrow R$$
$$f : x \longmapsto 6 + 3x + 10x^2 - 2x^4.$$

A graph of this function can be seen in Figure 10.1 as the solid line.

A majorizing function, g, is chosen as

$$g : \longmapsto 6 + 3x + 10x^2 - 8xy^2 + 6y^4.$$

The starting value for the algorithm is set at $x_0 = 1.4$, giving

$$g(x, 1.4) = 29.0496 - 18.952x + 10x^2,$$

a graph of which is shown in Figure 10.1 as the short dashed line. The mimimum of this quadratic function is easily found as $x = 0.948$. Hence $x_1 = 0.948$. The next iteration gives

$$g(x, 0.948) = 10.8460 - 3.7942x + 10x^2.$$

The graph of this function is shown as the long dashed line in Figure 10.1. The minimum of this function gives x_2, and the process continues until convergence at the minimum of f.

For metric MDS consider the loss function which will be called stress as

$$S = \sum_{r<s} w_{rs}(\delta_{rs} - d_{rs})^2, \tag{10.2}$$

where as usual $\{w_{rs}\}$ are weights, $\{\delta_{rs}\}$ are dissimilarities and $\{d_{rs}\}$ are Euclidean distances calculated from coordinates $\mathbf{X}$. Following Groenen (1993)

$$S = \sum_{r<s} w_{rs}\delta_{rs}^2 + \sum_{r<s} w_{rs}d_{rs}^2(\mathbf{X}) - 2\sum_{r<s} w_{rs}\delta_{rs}d_{rs}(\mathbf{X})$$
$$= \eta_\delta^2 + \eta^2(\mathbf{X}) - 2\rho(\mathbf{X}).$$

The stress S is now written in matrix form. Firstly

$$\eta^2(\mathbf{X}) = \sum_{r<s} w_{rs}(\mathbf{x}_r - \mathbf{x}_s)^T(\mathbf{x}_r - \mathbf{x}_s) = \text{tr}(\mathbf{X}^T\mathbf{V}\mathbf{X}),$$

where
$$[\mathbf{V}]_{rr} = \sum_{r \neq s} w_{rs} \qquad [\mathbf{V}]_{rs} = -w_{rs} \quad (r \neq s).$$

Next
$$\rho(\mathbf{X}) = \sum_{r<s} \frac{w_{rs}\delta_{rs}}{d_{rs}} d_{rs}^2$$
$$= \sum_{r<s} \frac{w_{rs}\delta_{rs}}{d_{rs}} (\mathbf{x}_r - \mathbf{x}_s)^T(\mathbf{x}_r - \mathbf{x}_s)$$
$$= \mathrm{tr}(\mathbf{X}^T\mathbf{B}(\mathbf{X})\mathbf{X}),$$

where
$$[\mathbf{B}(\mathbf{X})]_{rs} = w_{rs}\delta_{rs}/d_{rs}(\mathbf{X}) \quad \text{if } d_{rs}(\mathbf{X}) \neq 0,$$
$$= 0 \quad \text{if } d_{rs}(\mathbf{X}) = 0.$$

Then write stress as
$$S(\mathbf{X}) = \eta_\delta^2 + \mathrm{tr}(\mathbf{X}^T\mathbf{V}\mathbf{X}) - 2\mathrm{tr}(\mathbf{X}^T\mathbf{B}(\mathbf{X})\mathbf{X}).$$

A majorizing function, T, for stress S is given by
$$T(\mathbf{X}, \mathbf{Y}) = \eta_\delta^2 + \mathrm{tr}(\mathbf{X}^T\mathbf{V}\mathbf{X}) - 2\mathrm{tr}(\mathbf{X}^T\mathbf{B}(\mathbf{Y})\mathbf{Y}).$$

To show that T does majorize S,
$$\tfrac{1}{2}(T - S) = \rho(\mathbf{X}) - \tilde{\rho}(\mathbf{X}, \mathbf{Y}),$$

where
$$\tilde{\rho}(\mathbf{X}, \mathbf{Y}) = \mathrm{tr}(\mathbf{X}^T\mathbf{B}(\mathbf{Y})\mathbf{Y}).$$

Now
$$\rho(\mathbf{X}) = \sum_{r<s} w_{rs}\delta_{rs}\{(\mathbf{x}_r - \mathbf{x}_s)^T(\mathbf{x}_r - \mathbf{x}_s)\}^{\frac{1}{2}}$$
$$= \sum_{r<s} \frac{w_{rs}\delta_{rs}}{d_{rs}(\mathbf{Y})}\{(\mathbf{x}_r - \mathbf{x}_s)^T(\mathbf{x}_r - \mathbf{x}_s)(\mathbf{y}_r - \mathbf{y}_s)^T(\mathbf{y}_r - \mathbf{y}_s)\}^{\frac{1}{2}}$$
$$\geq \sum_{r<s} \frac{w_{rs}\delta_{rs}}{d_{rs}(\mathbf{Y})}(\mathbf{x}_r - \mathbf{x}_s)^T(\mathbf{y}_r - \mathbf{y}_s) = \tilde{\rho}(\mathbf{X}, \mathbf{Y})$$

by the Cauchy-Schwarz inequality, and hence $\rho(\mathbf{X}) \geq \tilde{\rho}(\mathbf{X}, \mathbf{Y})$. Also as $T(\mathbf{X}, \mathbf{X}) = S(\mathbf{X})$, T majorizes S.

To minimize T,
$$\frac{\partial T}{\partial \mathbf{Y}} = 2\mathbf{V}\mathbf{X} - 2\mathbf{B}(\mathbf{Y})\mathbf{Y} = \mathbf{0}. \tag{10.3}$$

Now $\mathbf{V}$ has rank $n - 1$ since its row sums are all zero, and so the Moore-Penrose inverse is used to solve equation (10.3), giving

$$\mathbf{X} = \mathbf{V}^+\mathbf{B}(\mathbf{Y})\mathbf{Y},$$

which is known as the Guttman transform as it appears in Guttman (1968).

Thus using the majorizing method for finding minimum stress simply has the Guttman transform as its updating equation. The algorithm gives rise to a non-decreasing sequence of stress values, which converge linearly (de Leeuw, 1988). One advantage of the majorizing method over gradient methods is that the sequence of stress values is always non-increasing. However it shares the same problem of not necessarily finding the global minimum, but can get stuck at a local minimum.

10.2.2 The majorizing method for nonmetric MDS

For nonmetric MDS the dissimilarities $\{\delta_{rs}\}$ are replaced by disparities $\{\hat{d}_{rs}\}$ in the loss function (10.2), and as with ALSCAL there are two minimizations to be carried out. In one the loss function or stress is minimized with respect to the distances $\{d_{rs}\}$, and in the other it is minimized with respect to the disparities $\{\hat{d}_{rs}\}$. The first minimization can be by the majorizing algorithm, the second by isotonic regression as discussed in Chapter 3.

10.2.3 Tunnelling for a global minimum

Groenen (1993) reviews methods for searching for global minima, and describes in detail the tunnelling method. Picturesquely, suppose you are at the lowest point of a valley in a mountainous region with only ascent possible in all directions. There is no direction in which you can descend, however you wish to be at a lower height above sea level. To overcome your predicament you dig horizontal tunnels in various directions through the surrounding mountains until from one tunnel you reach the other side of the mountain and descent is again possible.

The tunnelling method for stress first involves finding a configuration $\mathbf{X}^*$ which has local minimum stress. Then the tunnel is "dug" by finding other configurations with the same stress. The tunnelling function is defined as

$$\tau(\mathbf{X}) = \{S(\mathbf{X}) - S(\mathbf{X}^*)\}^{2\lambda} \left\{ 1 + \frac{1}{\sum_{rs}(d_{rs}(\mathbf{X}) - d_{rs}(\mathbf{X}^*))^2} \right\},$$

where λ is the pole strength parameter to be fixed, with $0 < \lambda < 1$. The zero points of $\tau(\mathbf{X})$ then give configurations which have the same stress as that for $\mathbf{X}^*$. The reader is referred to Groenen for details of how these zero points can be found. Once a new configuration is found the stress can then possibly be reduced further and hopefully a global minimum eventually reached.

De Leeuw (1977b) shows how the majorization method can be extended to general Minkowski spaces. Heiser (1991) shows how the method can be adapted to allow for some of the pseudo-distances being negative. The pseudo-distances are the quantities obtained from transformation of the dissimilarities. For example linear regression used on dissimilarities making them more distance-like could produce some negative pseudo-distances. Taking the method further, de Leeuw and Heiser (1980) show how the majorization method can be generalized to individual differences scaling.

CHAPTER 11

Further m-mode, n-way models

This chapter gives brief descriptions of some more MDS models appropriate for data of various numbers of modes and ways.

11.1 CANDECOMP, PARAFAC and CANDELINC

CANDECOMP (CANonical DEComposition) is a generalization of Carroll and Chang's (1970) INDSCAL model. The INDSCAL model, which is two-mode three-way, is written as

$$b_{rs,i} = \sum_{t=1}^{p} w_{it} x_{rt} x_{st}.$$

This can be generalized to the three-way CANDECOMP model for three-mode, three-way data,

$$z_{rsi} = \sum_{t=1}^{p} w_{it} x_{rt} y_{st}. \tag{11.1}$$

An example of three-mode, three-way data is where N judges of whisky, each rank m liquor qualities for each of n bottles of whisky. The model is fitted to data using a similar algorithm to the INDSCAL model. The least squares loss function for the three-way model can be written as

$$S = \sum_{i=1}^{N} ||\mathbf{Z}_i - \mathbf{X}\mathbf{D}_i\mathbf{Y}^T||^2, \tag{11.2}$$

where $\mathbf{Z}_i$ is the $n \times m$ matrix $[\mathbf{Z}]_{rs} = z_{rsi}$, $\mathbf{X}$ is the $n \times p$ matrix giving coordinates for the second mode of the data (bottles of whisky), $\mathbf{Y}$ is the $m \times p$ matrix giving coordinates for the third mode of the data (qualities), and $\mathbf{D}_i$ is a diagonal matrix for the first mode (judges).

The model (11.2) now corresponds to Harshman's (1970) PARA-FAC-1 (PARAllel profiles FACtor analysis) model, where $\mathbf{Z}_i$ is often viewed as a covariance matrix. The PARAFAC-2 model was seen as a special case of IDIOSCAL in Chapter 9. CANDECOMP can be used for data of more than three modes. Equation (11.1) can be generalized further to

$$z_{r_1 r_2 \ldots r_m} = \sum_{t=1}^{p} x_{r_1}^{(1)} x_{r_2}^{(2)} \ldots x_{r_m}^{(m)},$$

and can be fitted with a similar algorithm to that for the IND-SCAL. For further details see Carroll and Chang (1970), Harshman and Lundy (1984a,b), for an efficient algorithm for fitting the three-mode model, Kiers and Krijnen (1991), and for an alternating least squares algorithm, Kiers (1991).

Carroll *et al.* (1980) introduced CANDELINC (CANonical DEcomposition with LINear Constraints) which is the CANDECOMP model but incorporating constraints. The model is

$$z_{r_1 r_2 \ldots r_m} = \sum_{t=1}^{p} x_{r_1}^{(1)} x_{r_2}^{(2)} \ldots x_{r_m}^{(m)},$$

as before but with the constraints

$$\mathbf{X}_i = \mathbf{D}_i \mathbf{T}_i,$$

where $\mathbf{D}_i$ are known design matrices and $\mathbf{T}_i$ are matrices of unknown parameters. The design matrices, for example, can be used when dissimilarities are collected in an experiment according to some experimental design.

DEDICOM (DEcomposition into DIrectional COMponents) is a model devised by Harshman (1978) for analysing asymmetric data matrices. A one-mode two-way asymmetric $n \times n$ data matrix is decomposed as

$$\mathbf{X} = \mathbf{A}\mathbf{R}\mathbf{A}^T + \mathbf{N},$$

where $\mathbf{A}$ is an $n \times p$ matrix of weights $(p < n)$, $\mathbf{R}$ is a $p \times p$ matrix representing asymmetric relationships among the p dimensions, and $\mathbf{N}$ is an error matrix. The model can be fitted by using an alternating least squares algorithm; see Kiers (1989) and Kiers *et al.* (1990).

11.2 The Tucker models

Kroonenberg (1983) relates models in the previous section to the Tucker (1966) models, under the name of three-mode principal components analysis. A brief summary is given.

In Section 2.2.7 the connection between standard principal components analysis and classical scaling was discussed. The link between principal components analysis and the singular value decomposition of the data matrix forms the basis of the extension of principal components to three-way data. The sample covariance $n \times p$ matrix obtained from the mean-corrected data matrix, $\mathbf{X}$, is $(n-1)\mathbf{S} = \mathbf{X}^T\mathbf{X}$. Let the eigenvectors of $\mathbf{X}^T\mathbf{X}$ be $\mathbf{v}_i$ $(i = 1, \ldots, p)$, and placed in matrix $\mathbf{V}$. The component scores and component loadings are then given respectively by

$$\mathbf{XV} \quad \text{and} \quad \mathbf{V}.$$

However from Section 1.4.2 $\mathbf{V}$ is one of the orthonormal matrices in the singular value decomposition of $\mathbf{X}$,

$$\mathbf{X} = \mathbf{U}\Lambda\mathbf{V}^T.$$

Thus the component scores are given by

$$\mathbf{XV} = \mathbf{U}\Lambda\mathbf{V}^T\mathbf{V} = \mathbf{U}\Lambda.$$

Hence principal components analysis is equivalent to the singular value decomposition of the data matrix,

$$\mathbf{X} = \mathbf{U}\Lambda\mathbf{V}^T$$
$$= (\mathbf{U}\Lambda)\mathbf{V}^T$$
$$= \text{component scores} \times \text{component loadings}, \qquad (11.3)$$

and hence is an MDS technique since the component scores from the first few principal components are used to represent the objects or stimuli.

Write the singular value decomposition of $\mathbf{X}$ in (11.3) as

$$x_{ri} = \sum_{a=1}^{q} \sum_{b=1}^{q} u_{ra} v_{ib} \lambda_{ab} \qquad (r = 1, \ldots, n; i = 1, \ldots, p)$$

where $\mathbf{X}$ is of rank q. This gives the form of the generalization to three- or higher-mode data. The generalization to give three-mode principal components analysis is

$$z_{rst} = \sum_{i=1}^{I} \sum_{j=1}^{J} \sum_{k=1}^{K} u_{ri} v_{sj} w_{tk} \lambda_{ijk} \qquad (11.4)$$
$$(r = 1, \ldots, R; s = 1, \ldots, S; t = 1, \ldots, T),$$

where there has been a change in some of the notation.

The number of elements in the three modes are R, S and T respectively. The $R \times K$ matrix $\mathbf{U}$, where $[\mathbf{U}]_{ri} = u_{ri}$, contains the I "components" for the first mode, and similarly for matrices $\mathbf{V}$ and $\mathbf{W}$ for the second and third modes. These matrices are orthonormal, $\mathbf{U}^T \mathbf{U} = \mathbf{I}$, $\mathbf{V}^T \mathbf{V} = I$, $\mathbf{W}^T \mathbf{W} = \mathbf{I}$. The three-way $I \times J \times K$ matrix $[\mathbf{\Lambda}]_{ijk} = \lambda_{ijk}$ is the "core matrix" containing the relationships between various components.

The Tucker-1 model is standard principal components analysis on the three modes, using one pair of modes at a time. Equation (11.4) gives the Tucker-3 model where all three modes are equivalent in status, each having an orthonormal matrix of principal components, i.e. $\mathbf{U}$, $\mathbf{V}$, $\mathbf{W}$ respectively. The Tucker-2 model has $\mathbf{W}$ equal to the identity matrix, and so

$$z_{rst} = \sum_{i=1}^{I} \sum_{j=1}^{J} u_{ri} v_{sj} \lambda_{ijt},$$

giving the third mode special status, e.g. for judges ranking attributes on several objects.

A small number of components is desirable for each mode. The models are fitted using a least squares loss function. For the Tucker-3 model

$$\sum_{r=1}^{R} \sum_{s=1}^{S} \sum_{t=1}^{T} (z_{rst} - \hat{z}_{rst})^2$$

is minimized where

$$\hat{z}_{rst} = \sum_{i=1}^{I} \sum_{j=1}^{J} \sum_{k=1}^{K} u_{ri} v_{sj} w_{tk} \lambda_{ijk},$$

to give the estimated component matrices $\mathbf{U}$, $\mathbf{V}$, $\mathbf{W}$ and the core matrix $\mathbf{\Lambda}$. Kroonenberg (1983) discusses algorithms for fitting the Tucker models; see also Kroonenberg and de Leeuw (1980), and ten Berge et al. (1987) for an alternating least squares approach.

11.2.1 Relationship to other models

If the core matrix $\boldsymbol{\Lambda}$ is chosen as the three-way identity matrix then the Tucker-3 model becomes

$$z_{rst} = \sum_{i=1}^{I} u_{ri}v_{si}w_{ti},$$

which is equivalent to the PARAFAC-1 model, or the three-mode CANDECOMP model.

Let $\boldsymbol{\Lambda}$ be a three-way identity matrix and also let $\mathbf{U} \equiv \mathbf{V}$. Then the Tucker-3 model becomes

$$z_{rst} = \sum_{i=1}^{I} u_{ri}u_{si}w_{ti},$$

which is the INDSCAL model.

In the Tucker-2 model let $\mathbf{U} \equiv \mathbf{V}$, then

$$z_{rst} = \sum_{i=1}^{I} \sum_{j=1}^{J} u_{ri}u_{sj}\lambda_{ijt},$$

which is the IDIOSCAL model. The PARAFAC-2 model can then be obtained by making the off-diagonal elements of $\boldsymbol{\Lambda}$ equal.

11.3 One-mode, n-way models

Cox *et al.* (1992) consider a one-mode, n-way model. The model is best illustrated for the three-way case, where data are in the form of "three-way dissimilarities". Three-way dissimilarities $\{\delta_{rst}\}$ are generalized from two-way dissimilarities, so that δ_{rst} measures "how far apart" or "how dissimilar" the objects r, s and t are when considered as a triple. The requirement for δ_{rst} is that it is a real function such that

$$\delta_{rst} \geq 0 \qquad (r \neq s \neq t)$$

$$\delta_{rst} = \delta_{str} = \ldots = \delta_{\text{(every permutation of } r,s,t)} \qquad (r \neq s \neq t).$$

Dissimilarities δ_{rst} are only defined when r, s, t are distinct.

A configuration, $\mathbf{X}$, of points in a Euclidean space is sought that represents the objects, and a real-valued function, $d_{rst}(\mathbf{x}_r, \mathbf{x}_s, \mathbf{x}_t)$

constructed so that it satisfies the same conditions as the three-way dissimilarities. Possible functions are:

1. $d_{rst} = \max(d_{rs}, d_{rt}, d_{st})$, where d_{rs} is the Euclidean distance between the points r and s.
2. $d_{rst} = \min(d_{rs}, d_{rt}, d_{st})$.
3. $d_{rst} = (d_{rs}^2 + d_{rt}^2 + d_{st}^2)^{\frac{1}{2}}$.

For Euclidean distance between two points with coordinates $\mathbf{x}_r$ and $\mathbf{x}_s$,

$$d_{rs}^2 = \mathbf{x}_r^T \mathbf{x}_r + \mathbf{x}_s^T \mathbf{x}_s - \mathbf{x}_r^T \mathbf{x}_s.$$

If this is generalized to three points,

$$d_{rst}^2 = a(\mathbf{x}_r^T \mathbf{x}_r + \mathbf{x}_s^T \mathbf{x}_s + \mathbf{x}_t^T \mathbf{x}_t) + b(\mathbf{x}_r^T \mathbf{x}_s + \mathbf{x}_r^T \mathbf{x}_t + \mathbf{x}_s^T \mathbf{x}_t).$$

This function is symmetric in r, s, t. For invariance under rotation, translation and reflection it is easily shown that $a + b$ must equal zero. Choose $a = 2$ and $b = -2$, then

$$d_{rst}^2 = d_{rs}^2 + d_{rt}^2 + d_{st}^2.$$

This function is the one chosen to represent the three-way dissimilarities.

Stress is defined as

$$S = \left\{ \frac{\sum (d_{rst} - \hat{d}_{rst})^2}{\sum d_{rst}^2} \right\}^{\frac{1}{2}},$$

and can be fitted using a Kruskal type algorithm. Gradient terms can be found in Cox et $al.$ The fitting of this three-way model is denoted as MDS3. The extension to more than three ways, MDSn, is straightforward.

Cox et $al.$ argue the case for three-way scaling by calculating dissimilarities based on the Jaccard coefficient, s_{rs}, for the following data matrix consisting of seven binary variables recorded for four individuals.

$$\mathbf{X} = \begin{bmatrix} 1 & 0 & 0 & 1 & 1 & 0 & 0 \\ 0 & 0 & 0 & 0 & 1 & 1 & 1 \\ 0 & 1 & 1 & 0 & 1 & 0 & 0 \\ 1 & 0 & 1 & 0 & 0 & 0 & 0 \end{bmatrix}.$$

Define $\delta_{rs} = 1 - s_{rs}$, and then $\delta_{12} = \delta_{13} = \delta_{14} = \delta_{23} = \delta_{24} = \delta_{34} = \frac{4}{5}$, giving no discriminatory information about the four individuals. Define the three-way Jaccard coefficient, s_{rst}, as the number of

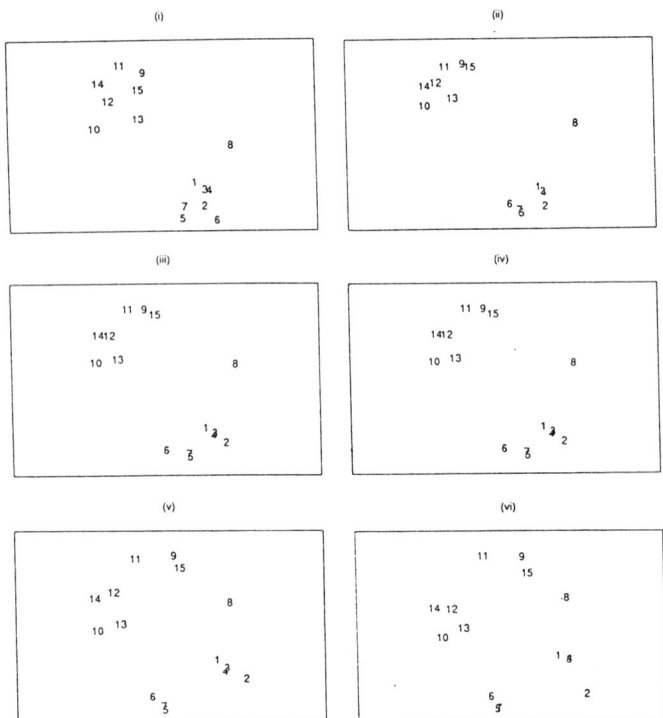

Figure 11.1 *MDS2 – MDS7 for the 1768 election data from Maidstone, Kent.*

variables in which individuals r, s and t each score "1" divided by the number of variables in which at least one of them scores "1". Let $\delta_{rst} = 1 - s_{rst}$. For the above data, $\delta_{123} = \frac{6}{7}$, $\delta_{124} = \delta_{134} = \delta_{234} = 1.0$, showing that the individuals 1, and 3 are in a group separated from individual 4. Cox *et al.* investigate one-mode, n-way scaling further on artificially constructed data, before applying it to some voting data from the 18th century in Great Britain.

Prior to the Ballot Act of 1872 electors had their votes recorded. They often had to vote for two, three or even more canditates. The data used were the votes cast on 20th June 1768 in the town of Maidstone in the county of Kent. There were fifteen candidates competing for seven available seats. The electors could vote for up to seven candidates.

The seven-way dissimilarity for a particular group of seven candidates was defined as the proportion of electors voting for the group subtracted from unity. MDS7 was then applied using a two dimensional space for the configuration. Results are shown in Figure 11.1(vi). The data can also be used to find 2,3,...,6-way dissimilarities, where for instance the dissimilarity among three particular candidates is the proportion of electors who voted for all three candidates among their seven votes subtracted from unity. Figures 11.1(i) to 11.1(v) show the configurations for MDS2 to MDS6 respectively. The stresses for the six configurations were 6%, 4%, 4%, 3%, 2%, and 1% respectively.

The six configurations in Figure 11.1 are similar to each other. MDS2 (standard nonmetric MDS) splits the candidates into two groups $\{1, 2, 3, 4, 5, 6, 7\}$ and $\{9, 10, 11, 12, 13, 14, 15\}$ together with a singleton $\{8\}$. However MDS2 does not reveal all the information about the voting behaviour of the electors. MDS5 shows candidate 6 moving towards candidate 8. MDS6 and MDS7 show candidate 7 also joining them.

A close look at the raw data shows that electors tended to vote for one or other of the two large groups in the MDS2 configuration. However those voting for the group $\{1, 2, 3, 4, 5, 6, 7\}$ tended to vote for candidates 1 to 5 and cast their other two votes mainly among candidates 6,7 and 8 but sometimes among candidates in the second group. The strong group 1 to 5 is confirmed in MDS5. Voters for the other large group usually voted for candidates 9 to 15 but with some votes going to other candidates, usually numbers 6,7 and 8. Voters often did not vote for both candidates 10 and 15, hence their separation in the configurations.

Pan and Harris (1991) consider a one-mode, n-way model. Let s_{rst} be a three-way similarity. Again a Euclidean space is sought in which points represent the objects. Let the coordinates be $\mathbf{x}_r$. Then the configuration is found such that

$$\lambda = \sum_{i=1}^{p} \lambda_i$$

is maximized, where

$$\lambda_i = -\frac{1}{2} \sum \sum_{r \neq s \neq t} \sum s_{rst} \frac{d_i^2(r,s,t)}{\sum_u x_{ui}^2},$$

$$d_i^2(r,s,t) = (x_{ri} - x_{si})^2 + (x_{ri} - x_{ti})^2 + (x_{si} - x_{ti})^2,$$

and $\mathbf{X}$ is constrained so that the centroid of the configuration is at the origin.

The problem can be written as the search for $\mathbf{X}$ that maximizes

$$\lambda = \sum_{i=1}^{p} \frac{\mathbf{x}_i^T \mathbf{H} \mathbf{x}_i}{\mathbf{x}_i^T \mathbf{x}_i},$$

subject to $\mathbf{X}^T \mathbf{1} = \mathbf{0}$, where

$$[\mathbf{H}]_{rs} = h_{rs} = \sum_{t, r \neq s \neq t} (s_{rst} + s_{str} + s_{trs}) \qquad (r \neq s)$$

$$= - \sum_{t, r \neq s \neq t} h_{rt} \qquad (r = s).$$

Pan and Harris use their model on some geological data from the Walker Lake quadrangle which includes parts of California and Nevada. Samples of stream sediments were analysed and elements measured (Fe, Mg, etc.). Similarity between triples of elements was measured by a generalization of the sample correlation coefficient. Interesting results occurred. One justification for using triples of elements rather than standard MDS for pairs, was that it was important to identify those elements which are closely associated with gold and silver together.

APPENDIX

Computer programs for multidimensional scaling

A.1 Computer programs

Early computer programs for nonmetric MDS were Kruskal's MD-SCAL, Guttman and Lingoes' smallest space analysis program SSA, and Young and Torgerson's TORSCA. The program MD-SCAL was later developed into KYST by Kruskal, Young and Seery. SSA was developed into MINISSA by Guttman, Lingoes and Roskam. Young also developed another program, POLYCON. Carroll and Chang introduced INDSCAL for individuals scaling, and later with Pruzansky, SINDSCAL. Later came Takane, Young and de Leeuw's ALSCAL, their alternating least squares program. Ramsay introduced MULTISCALE, a program for his maximum likelihood approach. SMACOF was developed by Heiser and de Leeuw.

Schiffman *et al.* (1981) discuss fully programs for MDS and the results obtained when they are used on various data sets. They also give addresses from where the programs can be obtained. In 1981 the MDS(X) series of programs was commercially launched, and included the following:

1. CANDECOMP – CANonical DECOMPosition;
2. HICLUS – HIerarchical CLUStering;
3. INDSCAL-S – INdividual Differences SCALing;
4. MDPREF – MultiDimensional PREFerence scaling;
5. MINICPA – Michigan-Israel-Nijmegen Integrated series; Conditional Proximity Analysis;
6. MINIRSA – MINI Rectangular Smallest space Analysis;
7. MINISSA – Michigan-Israel-Nijmegen Integrated Smallest Space Analysis;
8. MRSCAL – MetRic SCALing;
9. MVNDS – Maximum Variance Non-Dimensional Scaling;

10. PARAMAP – PARAmetric MAPping;
11. PINDIS – Procrustean INdividual DIfferences Scaling;
12. PREFMAP – PREFerence MAPping;
13. PROFIT – PROperty FITting;
14. TRIOSCAL – TRIadic similarities Ordinal SCALing;
15. UNICON – UNIdimensional CONjoint measurement.

The package is available from

Program Library Unit
University of Edinburgh
18 Buccleuch Place
Edinburgh, EH8 9LN
United Kingdom

It is now somewhat dated in its presentation.

More commonplace statistical packages which offer MDS are
SPSS(X) and SAS, both of which have an alternating least squares
option, ALSCAL. More limited are SYSTAT, STATISTICA and
SOLO which will carry out nonmetric MDS on a PC.

A.2 The accompanying diskette

The accompanying computer diskette contains programs, which
run under DOS, to carry out scaling techniques which have been
described in the text. The aim is to give the reader some "hands
on" experience of using the various techniques. However the condition of its use is that the authors and publisher do not assume
any liability for consequences from the use of the diskette. It is
suggested the files on the diskette are copied onto a hard disk
and run from there. For convenience the files can be put into a
directory (called MDS, say), a path set to the directory in the AU-
TOEXEC.BAT file, and then the programs run from a working
directory (MDSWORK). Note however, files containing data for
analysis must reside in the working directory. The programs can
be run by typing their name or from a menu invoked by typing
"MENU" at the keyboard. Various file names are asked for by the
programs for input and output of data. If a file name is given that
already exists, then the program repeats the request for the input of a file name. As programs are run, flags are placed in titles

within files in order to show the origins of various quantities. The techniques available are:

1. CLASSICAL SCALING;
2. NONMETRIC SCALING;
3. THREE-WAY NONMETRIC SCALING;
4. SPHERICAL SCALING;
5. PROCRUSTES ANALYSIS;
6. RECIPROCAL AVERAGING (CORRESPONDENCE ANALYSIS);
7. UNFOLDING;
8. INDIVIDUAL DIFFERENCES SCALING.

Also included on the diskette are the data sets used in the book, and various programs for manipulating dissimilarities, matrices and vectors. Two dimensional plots of points can also be accomplished with the program VEC_PLOT.

A.2.1 The programs

The following program displays a menu on the screen:

MENU — Aimed to provide a menu structure of available programs.

The programs which carry out the scaling techniques are:

CLSCAL — Classical scaling.

INDSCAL — Individual differences scaling.

MDSCAL_2 — Program uses method of Kruskal (plus Lingoes and Roskam) to produce iteratively an acceptable configuration of points in low dimension starting from a set of dissimilarities. Modified to consider a two subscript version $d(i, j)$.

MDSCAL_3 — Program uses method of Kruskal (plus Lingoes and Roskam) to produce iteratively an acceptable configuration of points in low dimension starting from a set of dissimilarities. Modified to consider a three subscript version $d(i, j, k)$.

MDSCAL_T — Program uses method of Kruskal (plus Lingoes and Roskam) to produce iteratively an acceptable configuration of points in low dimension starting from a set of dissimilarities. Modified to consider a two subscript version $d(i, j)$ on a sphere.

PROCRUST — Procrustes fits configuration with overlap of points.

RECIP_1 — Program to carry out the reciprocal averaging technique, now also known as correspondence analysis. Optimized version to pose the problem as a minimum size eigenmatrix. Set up to deal with the special case of few rows and many columns.

RECIPEIG — Program to carry out the reciprocal averaging technique, now also known as correspondence analysis. Optimized version to pose the problem as a minimum size eigenmatrix.

UNFOLDIN — Metric unfolding.

The following programs plot configurations on the screen:

VEC_PLOT — Vector plotting package (Only two dimensional plots are available).

SHEP_PLO — Shepard plot.

THETA_PL — Three dimensional plotting program for spherical data.

The remaining programs deal with data and dissimilarities:

CON2UNFI — Converts an integer contingency table (as in RECIPEIG) to one appropriate for unfolding.

CON2UNFR — Converts a real contingency table (data matrix) to one appropriate for unfolding. Copes with format length problems imposed by DOS.

HIST — Converts a matrix of historical voting data to give coefficients for input to the two or three multidimensional scaling programs.

IND2CON — Program to transform an indicator matrix to a contingency table.

MAT_COR2 — Converts a data matrix to give a correlation matrix for input to the multidimensional scaling programs. Employs the square root of $(1 - \text{correlation})/2$.

MAT2JACC — Converts an individual versus attribute data matrix to give Jaccard coefficients for input to the multidimensional scaling programs.

MDS_INPU — Prompts for input of dissimilarity data and outputs in a format and order required for the multidimensional scaling program in two or three dimensions.

TRANSPOS — Transposes an information matrix, swapping the individual/attribute roles.

UNF_JOIN — Combines two unfolding vectors into a single vector with separate colours.

VEC2CSV — Converts vectors into comma separated values suitable for importing into a spread sheet.

VEC_DISS — Generates two- or three-way dissimilarity data to match a specific vector useful in verifying the programs.

VEC2GOWE — Generates two-way dissimilarity data to match a specific vector using Gower's general dissimilarity.

A.2.2 The data sets

UK_TRAVE.DIS (Chapter 1)

SKULLS.DAT (Chapter 2)

KELLOG.DAT (Chapter 2, Chapter 3)

WORLD_TR.DAT (Chapter 3, Chapter 4)

WORLD_TR.DEG (Chapter 4)

HANS_70.DAT HANS_71.DAT HANS_72.DAT HANS_73.DAT (Chapter 4)

ORD_SURV.VEC SPEED.VEC (Chapter 5)

MONK_84.DIS MONK_85.DIS (Chapter 6)

AIR_EXPE.VEC AIR_NOVI.VEC (Chapter 6)

YOGHURT.DAT (Chapter 6)

NATIONS.DAT (Chapter 7)

CANCER.DAT (Chapter 8)

MUNSINGE.DAT (Chapter 8)

BIRTH.IND (Chapter 8)

PGEB.DAT PGWC.DAT POQ.DAT TPO.DAT TRD.DAT
(Chapter 9)

MAIDSTONE.68 (Chapter 11)

A.2.3 An illustration

The following commands will recreate the configuration in Figure
2.2 – nonmetric scaling of the breakfast cereal data.

VEC2GOWE
KELLOG.DAT (The file containing the data)
ECHO1 (A file to echo screen results)
DISSIM (A file in which to store the dissimilarities)
MDSCAL_2
DISSIM (The input file of dissimilarities)
ECHO2 (A file to echo screen results)
(return) (No input of starting configuration)
2 (Number of dimensions)
100 (The number of iterations)
CONFIG (A file to store the final configuration coordinates)
SHEP (A file to store data for a Shepard plot)
VEC_PLOT
CONFIG (The configuration to plot)
SHEP_PLO
SHEP (Data for a Shepard plot)

Alternatively the MENU can be used by typing "MENU" and
then the appropriate number referring to the particular procedures.
"MENU" has to be retyped after each procedure has finished ex-
ecuting.

Figures in the text
The configurations in the text can be reproduced using the following programs and data sets

Figure 1.1	CLSCAL	UK_TRAVE.DIS
	VEC_PLOT	
Figure 2.1	VEC_DISS	SKULLS.DAT
	CLSCAL	
	VEC_PLOT	
Figure 2.2	VEC2GOWE	KELLOG.DAT

	MDSCAL_2	
Figure 3.2	VEC2GOWE	KELLOG.DAT
and	MDSCAL_2	
Figure 3.4	VEC_PLOT	
	SHEP_PLO	
Figure 3.11	VEC2GOWE	KELLOG.DAT
	MDSCAL_2	
Figure 3.12	MAT2JACC	WORLD_TR.DAT
	MDSCAL_2	
	VEC_PLOT	
Figure 4.2	VEC_DISS	HANS_70.DAT
	MDSCAL_2	HANS_71.DAT
	PROCRUST	HANS_72.DAT
		HANS_73.DAT
Figure 4.3	MAT2JACC	WORLD_TR.DAT
	MDSCAL_T	WORLD_TR.DEG
	THETA_PL	
Figure 5.2(ii)	PROCRUST	ORD_SURV.VEC
	VEC_PLOT	SPEED.VEC
Figure 6.1(i)	MDSCAL_2	MONK_84.DIS
	VEC_PLOT	
Figure 6.1(ii)	MDSCAL_2	MONK_85.DAT
	PROCRUST	
	VEC_PLOT	
Figure 6.2(iii)	PROCRUST	AIR_EXPE.VEC
	VEC_PLOT	AIR_NOVI.VEC
Figure 6.3	VEC_DISS	YOGHURT.DAT
	MDSCAL_2	
	VEC_PLOT	
Figure 7.1	CON2UNFR	NATIONS.DAT
	UNFOLDIN	
	UNF_JOIN	
	VEC_PLOT	
Figure 8.2	TRANSPOS	CANCER.DAT
	RECIPEIG	
	VEC_PLOT	

A.3 Data format

Vectors need to be in the following (FORTRAN) format:

	FORMAT
Heading	**A80**
I,J,K,ALPHA	**3I3,G16.9**
ACR,COL,X	**2A4,5G16.9**
(for i=1,...,I)	

where: I is number of individuals; J is number of dimensions for the solution; K is number of cycles performed in the analysis; AL-PHA is final gradient calculated; ACR is an identifier (acronym) for individual i; COL is a descriptor (colour) for individual i; X is the vector position of individual i.

Note this structure is designed for use with multidimensional scaling. However it is adopted for all the programs. In general the parameters J, K and ALPHA may be omitted.

Dissimilarities need to be in the following FORMAT:

	FORMAT
Heading	**A80**
I	**I3**
dij	**G16.9**
(for i=2,...,I;	
j=1,...,i-1)	
ACR,COL	**2A4**
(for i=1,...,I)	

where: I is the number of individuals; dij is the dissimilarity between individuals i and j (assign missing values a negative dissimilarity); ACR is an identifier (acronym) for individual i; COL is a descriptor (colour) for individual i.

Note that the dissimilarities are input by rows. Since the matrix is symmetric, only the lower triangle is required. In general the parameters ACR and COL may be omitted. If required successive integers will be adopted to label the points.

Dissimilarities for individual differences scaling

The format closely follows that of the dissimilarity data files.

	FORMAT
Heading	**A80**
I,J	**2I3**
dijk	**G16.9**
(for i=1,...,I;	
j=2,...,J;	
k=1,...,j-1)	
ACR,COL	**2A4**
(for i=1,...,I)	
ACR1,COL1	**2A4**
(for j=1,...,J)	

where: I is the number of individuals; J is the number of dissimilarity matrices; dijk is the dissimilarity for individual i between objects j and k (assign missing values a negative dissimilarity); ACR is an identifier (acronym) for individual i; COL is a descriptor (colour) for individual i; ACR1 is an identifier (acronym) for object j; COL1 is a descriptor (colour) for object j.

Note that the dissimilarities are input by rows. Since the matrix is symmetric, only the lower triangle is required. Missing dissimilarities are unacceptable for this technique. In general the parameters ACR, COL and ARC1, COL1 may be omitted. If required successive integers will be adopted to label the points.

Contingency tables

The following is for contingency tables or data matrices with integer values.

	FORMAT
Heading	**A80**
I,J	**2I3**
A	**80I3**
(for i=1,...,I)	
ACR,COL	**2A4**
(for i=1,...,I)	
ACR1,COL1	**2A4**
(for j=1,...,J)	

where: I is the number of individuals (rows); J is the number of attributes (columns); A is the J dimensional row vector of attributes for individual i; ACR is an identifier (acronym) for individual i; COL is a descriptor (colour) for individual i; ACR1 is an identifier (acronym) for attribute j; COL1 is a descriptor (colour) for attribute j.

In general the parameters ACR, COl and ARC1, COL1 may be omitted. If required successive integers will be adopted to label the points.

Indicator matrix

This "indicator matrix" stores the values in a contingency table in compact form for use in multiple correspondence analysis. For example, from page 138, the 28 rows of (0, 0, 0, 1, 0, 0, 1) would have a single line entry in the matrix as 28 4 3.

Heading	**A80**
Frequency, Levels	**11I3**
(for i=1,...,I)	
ACR,COL	**2A4**
(for i=1,...,I)	

where: I is the number of levels; ACR is an identifier (acronym) for individual i; COL is a descriptor (colour) for individual i.

In general the parameters ACR and COL may be omitted. If required successive integers will be adopted to label the points.

A.4 Error messages

All file allocations and parameters specific to the programs are set interactively at run time. In particular a file containing a record of the run is compiled. Appropriate data files as described above are prepared in advance.

Any errors associated with the run, which typically arise if too large a data set is considered, will be reported on the screen. In addition stopping codes are produced; a value of 0 (STOP 0) is associated with a successful run. The other stopping codes are summarized below.

CLSCAL

1 - No file of dissimilarities provided
2 - No file to record the output provided
3 - No file to record the final configuration provided
4 - Too many individuals required
5 - A missing value encountered in the dissimilarity list
6 - Increase boundary arrays required for the sort subroutine
7 - Insufficient storage for temporary arrays in sort

HIST

1 - Too many candidates - array overload
2 - Too many pairings of candidates - array overload

INDSCAL

1 - No file of dissimilarities provided
2 - No file to record the output provided
3 - No file to record the final configuration provided
4 - No file to record the attribute weights
5 - Too many individuals required
6 - Too many attributes required
7 - A missing value encountered in the dissimilarity list

IND2INF

1 - No file with an indicator matrix provided
2 - Insufficient storage for the input vectors
3 - Insufficient storage for the grand total
4 - Insufficient storage for the Burt matrix
5 - No file to record the information matrix

INF2UNFI

1 - No file with an information matrix provided
2 - No output file for the unfolding matrix
3 - Insufficient storage for the data
4 - A missing value encountered in the data

INF2UNFR

1 - No file with an information matrix provided
2 - No output file for the unfolding matrix
3 - Insufficient storage for the data
4 - A missing value encountered in the data

MAT_COR2

1 - No file to record the output provided
2 - No output file for the generated dissimilarities

Note that the program will cycle until no more data sets are
provided.

MAT2JACC

1 - Insufficient storage provided for the local variables
2 - No information matrix provided
3 - No file to record the output provided
4 - No file to record the dissimilarities for the individuals provided
5 - No file to record the dissimilarities for the attributes provided
6 - Too many individuals required
7 - Too many attributes required

MDSCAL_T

1 - No file of dissimilarities provided

2 - No file to record the output provided
3 - Too many cycles required in a single run
4 - No file to record the final configuration provided
5 - No file to record the results for a Shepherd plot
6 - Too many individuals required
7 - Increase boundary arrays required for the sort subroutine

Note that typically 100 cycles will suffice.

MDSCAL_2

1 - No file of dissimilarities provided
2 - No file to record the output provided
3 - Too many cycles required in a single run
4 - No file to record the final configuration provided
5 - No file to record the results for a Shepherd plot
6 - Too many individuals required
7 - The solution is required in too many dimensions
8 - Increase boundary arrays required for the sort subroutine

Note that typically 100 cycles will suffice for a two dimensional solution.

MDSCAL_3

1 - No file of dissimilarities provided
2 - No file to record the output provided
3 - Too many cycles required in a single run
4 - No file to record the final configuration provided
5 - No file to record the results for a Shepherd plot
6 - Too many individuals required
7 - The solution is required in too many dimensions
8 - Increase boundary arrays required for the sort subroutine

Note that typically 100 cycles will suffice for a two dimensional solution.

MDS_INPU

1 - No file to record the output provided
2 - No file to record the dissimilarities provided

PROCRUST

1 - No file to record the output provided
2 - No mobile configuration file provided
3 - No file to record the new mobile configuration provided
4 - Too many points in the target cluster
5 - Too many dimensions required for the target cluster
6 - Too many points in the mobile cluster
7 - Too many dimensions required for the mobile cluster
8 - No common points in the two clusters - Are the acronyms correct?
9 - Increase boundary arrays required for the sort subroutine
10 - Insufficient storage for temporary arrays in sort

The program will cycle until no more target configurations are provided.

RECIP_1

1 - No file provided for the information matrix
2 - No file to record the output provided
3 - No file to record the final configuration provided
4 - Too many individuals or attributes required
5 - Bounds exceeded for the eigen problem
6 - The data contains an excess of missing values
7 - Increase boundary arrays required for the sort subroutine
8 - Insufficient storage for temporary arrays in sort

SHEP_PLO

1 - No file provided for the input vector

RECIPEIG

1 - No file provided for the information matrix
2 - No file to record the output provided
3 - No file to record the final configuration provided
4 - Too many individuals or attributes required
5 - Bounds exceeded for the eigen problem
6 - The data contains an excess of missing values
7 - Increase boundary arrays required for the sort subroutine
8 - Insufficient storage for temporary arrays in sort

THETA_PL

1 - No file provided for the input vector
2 - No file to record the output required
3 - All translations, scalings and rotations zero

While not reported as an error any overlap which occurs when plotting acronyms has the offending characters replaced by a star. Note that the vector for plotting is in radians as generated by MDSCAL_T, while any target vector is in degrees for ease of input.

TRANSPOS

1 - Too many individuals required
2 - Too many attributes required
3 - No file provided for the information matrix

The program will cycle until no more information matrices are available.

UNF_JOIN

1 - No file provided for the second vector
2 - No file to record the final configuration provided

The program will cycle until no more vectors are available.

UNFOLDIN

1 - No file of dissimilarities provided
2 - No file to record the output required
3 - No file provided to record the X vector
4 - No file provided to record the Y vector
5 - Insufficient space for the X dimension
6 - Insufficient space for the Y dimension
7 - Insufficient space for the dimensions required
8 - Increase boundary arrays required for the sort subroutine
9 - Insufficient storage for temporary arrays in sort
10 - A null vector has been generated

VEC_PLOT

1 - No file provided for the input vector

While not reported as an error any overlap which occurs when plotting acronyms has the offending characters replaced by a star.

VEC2CSV

1 - No file provided to record the output vector

The program will cycle until no more vectors are available.

VEC_DISS

1 - No vector file provided
2 - No file to record the output provided
3 - No file to record the dissimilarities provided
4 - Too many individuals required
5 - Too many dimensions required

VEC2GOWE

1 - No vector file provided
2 - No file to record the output provided
3 - No file to record the dissimilarities provided
4 - Too many individuals required
5 - Too many dimensions required

References

Ambrosi, K. and Hansohm, J. (1987) Ein dynamischer Ansatz zur Repräsentation von Objekten. In *Operations Research Proceedings 1986*, Berlin: Springer-Verlag.

Anderberg, M.R. (1973) *Cluster Analysis for Applications*, New York: Academic Press.

Anderson, A.J.B. (1971) Ordination methods in ecology. *J. of Ecol.*, **59**, 713-726.

Andrews, D.F. and Herzberg, A.M. (1985) *Data*, New York: Springer-Verlag.

Backhaus, W., Menzel, R. and Kreissl, S. (1987) Multidimensional scaling of color similarity in bees. *Biol. Cybern.*, **56**, 293-304.

Barlow, R.E., Bartholomew, D.J., Bremner, J.M. and Brunk, H.D. (1972) *Statistical Inference under Order Restrictions*. London: Wiley.

Barnett, S. (1990) *Matrices: Methods and Applications*, Oxford: Oxford University Press.

Bénasséni, J. (1993) Perturbational aspects in correspondence analysis. *Computational Statistics & Data Analysis*, **15**, 393-410.

Bennett, J.F. and Hays, W.L. (1960) Multidimensional unfolding: determining the dimensionality of ranked preference data. *Psychometrika*, **25**, 27-43.

Bennett, J.M. (1987) Influential observations in multidimensional scaling. In Heiberger, R.M. (ed.), *Proceedings of the 19th Symposium on the Interface (Computer Science and Statistics)*, Am. Stat. Assoc., pp 147-154.

Bentler, P.M. and Weeks, D.G. (1978) Restricted multidimensional scaling models. *Journal of Mathematical Psychology*, **17**, 138-151.

Benzécri, J.P. (1992) *Correspondence Analysis Handbook*, New York: Marcel Dekker.

Bloxom, B. (1978) Constrained multidimensional scaling in *N* spaces. *Psychometrika*, **43**, 283-319.

Borg, I. (1977) Geometric representation of individual differences. In Lingoes, J.C. (ed.), *Geometric Representations of Relational Data*, Ann Arbor, Mich: Mathesis.

Borg, I. and Lingoes, J.C. (1980) A model and algorithm for multidimensional scaling with external constraints on the distances. *Psychometrika*, **45**, 25-38.

Brady, H.E. (1985) Statistical consistency and hypothesis testing for nonmetric multidimensional scaling. *Psychometrika*, **50**, 509-537.

Bricker, C., Tooley, R.V. and Crone, G.R. (1976) *Landmarks of Mapmaking: An Illustrated Survey of Maps and Mapmakers*, New York: Thomas Y Cromwell.

Brokken, F.B. (1983) Orthogonal Procrustes rotation maximizing congruence. *Psychometrika*, **48**, 343-349.

Browne, M.W. (1967) On oblique Procrustes rotation. *Psychometrika*, **32**, 125-132.

Büyükkurt, B.K. and Büyükkurt, M.D. (1990) Robustness and small-sample properties of the estimators of probabilistic multidimensional scaling (PROSCAL). *Journal of Marketing Research*, **27**, 139-149.

Cailliez, F. (1983) The analytical solution of the additive constant problem. *Psychometrika*, **48**, 305–308.

Carroll, J.D., and Arabie, P. (1980) Multidimensional scaling. *Ann. Rev. Psychol.*, **31**, 607-649.

Carroll, J.D. and Chang, J.J (1970) Analysis of individual differences in multidimensional scaling via an n-way generalization of "Eckart-Young" decomposition. *Psychometrika*, **35**, 283-319.

Carroll, J.D. and Chang, J.J. (1972) IDIOSCAL (Individual Differences in Orientation Scaling): a generalization of INDSCAL allowing idiosyncratic references systems. Paper presented at Psychometric Meeting, Princeton, NJ.

Carroll, J.D. and Wish, M. (1974) Models and methods for three-way multidimensional scaling. In Krantz, D.H, Atkinson, R.L., Luce, R.D. and Suppes, P. (eds.), *Contemporary Developments in Mathematical Psychology, Vol. 2*, San Francisco: W.H. Freeman.

Carroll, J.D., Pruzansky, S. and Kruskal, J.B. (1980) CANDELINC: A general approach to multidimensional analysis of many-way arrays with linear constraints on parameters. *Psychometrika*, **45**, 3-24.

Chang, C.L. and Lee, R.C.T. (1973) A heuristic relaxation method for non-linear mapping in cluster analysis. *I.E.E.E. Trans. on Systems, Man. and Cybernetics*, **3**, 197-200.

Chatfield, C. and Collins, A.J. (1980) *Introduction to Multivariate Analysis*, London: Chapman and Hall.

Choulakian, V. (1988) Exploratory analysis of contingency tables by log-linear formulation and generalizations of correspondence analysis. *Psychometrika*, **53**, 235-250.

Cliff, N. (1968) The "idealized individual" interpretation of individual differences in multidimensional scaling. *Psychometrika*, **33**, 225-232.

Cliff, N., Girard, R., Green, R.S., Kehoe, J.F. and Doherty, L.M. (1977)

INTERSCAL: A TSO FORTRAN IV program for subject computer interactive multidimensional scaling. *Educational and Psychological Measurement*, **37**, 185-188.

Coombs, C.H. (1950) Psychological scaling without a unit of measurement. *Psychol. Rev.*, **57**, 148-158.

Coombs, C.H. (1964) *A Theory of Data*, New York: Wiley.

Coombs, C.H. and Kao, R.C. (1960) On a connection between factor analysis and multidimensional unfolding. *Psychometrika*, **25**, 219-231.

Cooper, L.G. (1972) A new solution to the additive constant problem in metric multidimensional scaling. *Psychometrika*, **37**, 311-321.

Cormack, R.M. (1971) A review of classification (with Discussion). *J. R. Statist. Soc.*, **A 134**, 321-367.

Corradino, C. (1990) Proximity structure in a captive colony of Japanese monkeys (*Macaca fuscata fuscata*): an application of miltidimensional scaling. *Primates*, **31**, 351-362.

Coury, B.G. (1987) Multidimensional scaling as a method of assessing internal conceptual models of inspection tasks. *Ergonomics*, **30**, 959-973.

Cox, M.A.A. and Cox, T.F. (1992) Interpretation of stress in nonmetric multidimensional scaling. *Statistica Applicata*, **4**, 611-618.

Cox, T.F. and Cox, M.A.A. (1990) Interpreting stress in multidimensional scaling. *J. Statist. Comput. Simul.*, **37**, 211-223.

Cox, T.F. and Cox, M.A.A. (1991) Multidimensional scaling on a sphere. *Commun. Statist.*, **20**, 2943-2953.

Cox, T.F., Cox, M.A.A. and Branco, J.A. (1992) Multidimensional scaling for n-tuples. *British Journal of Mathematical and Statistical Psychology*, **44**, 195-206.

Cramer, E.M. (1974) On Browne's solution for oblique Procrustes rotation. *Psychometrika*, **39**, 159-163.

Critchley, F. (1978) Multidimensional scaling: a short critique and a new method. In Corsten, L.C.A. and Hermans, J. (eds.), *COMPSTAT 1978*, Vienna: Physica-Verlag.

Davidson, J.A. (1972) A geometrical analysis of the unfolding model: nondegenerate solutions. *Psychometrika*, **37**, 193-216.

Davidson, J.A. (1973) A geometrical analysis of the unfolding model: general solutions. *Psychometrika*, **38**, 305-336.

Davidson, M.L. (1983) *Multidimensional scaling*, New York: Wiley.

Davies, P.M. and Coxon, A.P.M. (1983) *The MDS(X) User Manual*, University of Edinburgh, Program Library Unit.

De Leeuw, J. (1977a) Correctness of Kruskal's algorithms for monotone regression with ties. *Psychometrika*, **42**, 141-144.

De Leeuw, J.A. (1977b) Applications of convex analysis to multidimensional scaling. In Barra, J.R., Brodeau, F., Romier, G. and van Cutsen,

B. (eds.), *Recent Developments in Statistics*, Amsterdam: North Holland, pp 133-145.

De Leeuw, J. (1984) Differentiability of Kruskal's stress at a local minimum. *Psychometrika*, **49**, 111-113.

De Leeuw, J. (1988) Convergence of the majorization method for multidimensional scaling. *Journal of Classification*, **5**, 163-180.

De Leeuw, J. (1992) Fitting distances by least squares. Unpublished report.

De Leeuw, J. and Heiser, W. (1977) Convergence of correction matrix algorithms for multidimensional scaling. In Lingoes, J.C. (ed.), *Geometric Representations of Relational Data*, Ann Arbor, Michigan: Mathesis Press.

De Leeuw, J. and Heiser, W. (1980) Multidimensional scaling with restrictions on the configuration. In Krishnaiah, P.R. (ed.), *Multivariate Analysis V*, Amsterdam: North Holland.

De Leeuw, J. and Heiser, W. (1982) Theory of multidimensional scaling. In Krishnaiah, P.R. and Kanal, L.N. (eds.), *Handbook of Statistics, Vol 2*, Amsterdam: North Holland, pp 285-316.

De Leeuw, J. and Stoop, I. (1984) Upper bounds for Kruskal's stress. *Psychometrika*, **49**, 391-402.

De Leeuw, J. and van der Heijden, P.G.M. (1988) Correspondence analysis of incomplete contingency tables. *Psychometrika*, **53**, 223-233.

De Leeuw, J., Young, F.W. and Takane, Y. (1976) Additive structure in qualitative data: an alternating least squares method with optimal scaling features. *Psychometrika*, **41**, 471-503.

Diday, E. and Simon, J.C. (1976) Clustering analysis. In Fu, K.S. (ed.), *Communication and Cybernetics 10 Digital Pattern Recognition*, Berlin: Springer-Verlag.

Digby, P.G.N. and Kempton, R.A. (1987) *Multivariate Analysis of Ecological Communities*. London: Chapman and Hall.

Diggle, P.J. (1983) *Statistical Analysis of Spatial Point Patterns*. London: Academic Press.

Dobson, A.J. (1983) *Introduction to Statistical Modelling*. London: Chapman and Hall.

Ekman, G. (1954) Dimensions of colour vision. *Journal of Psychology*, **38**, 467-474.

Fagot, R.F. and Mazo, R.M. (1989) Association coefficients of identity and proportionality for metric scales. *Psychometrika*, **54**, 93-104.

Fawcett, C.D. (1901) A second study of the variation and correlation of the human skull, with special reference to the Naqada crania. *Biometrika*, **1**, 408-467.

Fenton, M. and Pearce, P. (1988) Multidimensional scaling and tourism research. *Annals of Tourism Research*, **15**, 236-254.

Fisher, R.A. (1940) The precision of discriminant functions. *Ann. Eugen.*, **10**, 422-429.

Fitzgerald, L.F. and Hubert, L.J. (1987) Multidimensional scaling: some possibilities for counseling psychology. *Journal of Counseling Psychology*, **34**, 469-480.

Gilula, Z. and Ritov, Y. (1990) Inferential ordinal correspondence analysis: motivation, derivation and limitations. *International Statistical Review*, **58**, 99-108.

Girard, R.A. and Cliff, N. (1976) A Monte Carlo evaluation of interactive multidimensional scaling. *Psychometrika*, **41**, 43-64.

Gold, E.M. (1973) Metric unfolding: data requirement for unique solution and clarification of Schönemann's algorithm. *Psychometrika*, **38**, 555-569.

Goodall, D.W. (1967) The distribution of the matching coefficient. *Biometrics*, **23**, 647-656.

Gordon, A.D. (1981a) *Classification.* London: Chapman and Hall.

Gordon, A.D. (1981b) Constructing dissimilarity measures. *Journal of Classification*, **7**, 257-269.

Gower, J.C. (1966) Some distance properties of latent root and vector methods in multivariate analysis. *Biometrika*, **53**, 325-338.

Gower, J.C. (1971) A general coefficient of similarity and some of its properties. *Biometrics*, **27**, 857-874.

Gower, J.C. (1975) Generalized Procrustes analysis. *Psychometrika*, **40**, 33-51.

Gower, J.C. (1977) The analysis of asymmetry and orthogonality. In Barra, J.R. et al. (eds.), *Recent Developments in Statistics*, Amsterdam: North Holland.

Gower, J.C. (1984) Multivariate analysis: ordination, multidimensional scaling and allied topics. In Lloyd, E.H. (ed.), *Handbook of Applicable Mathematics*, vol VI, New York: Wiley.

Gower, J.C. (1985) Measures of similarity, dissimilarity and distance. In Kotz, S., Johnson, N.L. and Read, C.B. (eds.), *Encyclopedia of Statistical Sciences, Vol 5*, pp 397-405.

Gower, J.C. (1990) Fisher's optimal scores and multiple correspondence analysis. *Biometrics*, **46**, 947-961.

Gower, J.C. and Legendre, P. (1986) Metric and Euclidean properties of dissimilarity coefficients. *Journal of Classification*, **3**, 5-48.

Green, B.F. (1952) The orthogonal approximation of an oblique structure in factor analysis. *Psychometrika*, **17**, 429-440.

Green, P.J. and Sibson, R. (1978) Computing Dirichlet tesselations in the plane. *Computer J.*, **21**, 168-173.

Green, R.S. and Bentler, P.M. (1979) Improving the efficiency and effectiveness of interactively selected MDS data designs. *Psychometrika*, **44**, 115-119.

Greenacre, M.J. (1984) *Theory and Applications of Correspondence Analysis*, London: Academic Press Inc.

Greenacre, M.J. (1988) Correspondence analysis of multivariate categorical data by weighted least-squares. *Biometrika*, **75**, 457-467.

Greenacre, M.J. and Browne, M.W. (1986) An efficient alternating least-squares algorithm to perform multidimensional unfolding. *Psychometrika*, **51**, 241-250.

Greenacre, M.J. and Hastie, T. (1987) The geometrical interpretation of correspondence analysis. *JASA*, **82**, 437-447.

Greenacre, M.J., and Underhill, L.G. (1982) Scaling a data natrix in a low dimensional Euclidean space. In Hawkins, D.M. (ed.), *Topics in Applied Multivariate Analysis*, Cambridge: Cambridge University Press, pp 183-268.

Groenen, P.J.F. (1993) *The Majorization Approach to Multidimensional Scaling: Some Problems and Extensions*. Leiden: DSWO Press.

Guttman, L. (1941) The quantification of a class of attributes: a theory and method of scale construction. In Horst, P. *et al.* (eds.), *The Prediction of Personal Adjustment*, New York: Social Science Research Council, pp. 319-348.

Guttman, L. (1968) A general nonmetric technique for finding the smallest coordinate space for a configuration of points. *Psychometrika*, **33**, 469-506.

Hansohm, J. (1987) DMDS dynamic multidimensional scaling. Report, University of Augsburg.

Harshman, R.A. (1970) Foundations of the PARAFAC procedure: models and conditions for an "explanatory" multi-mode factor analysis. UCLA Working Papers in Phonetics, **16**.

Harshman, R.A. (1972) Determination and proof of minimum uniqueness conditions for PARAFAC-1. UCLA Working Papers in Phonetics , **22**.

Harshman, R.A. (1978) Models for analysis of asymmetrical relationships among *N* objects or stimuli. Paper presented at the First Joint Meeting of the Psychometric Society and the Society of Mathematical Psychology, Hamilton, Ontario.

Harshman, R.A. and Lundy, M.E. (1984a) The PARAFAC model for three-way factor analysis and multidimensional scaling. In Law, H.G., Snyder, C.W., Hattie, J.A. and McDonald, R.P. (eds.), *Research Methods for Multimode Data Analysis*. New York: Praeger, pp 122-215.

Harshman, R.A. and Lundy, M.E. (1984b) Data preprocessing and extended PARAFAC model. In Law, H.G., Snyder, C.W., Hattie, J.A. and McDonald, R.P. (eds.), *Research Methods for Multimode Data Analysis*. New York: Praeger, pp 216-284.

Hartigan, J.A. (1967) Representation of similarity matrices by trees. *J. Am. Statist. Ass.*, **62**, 1140-1158.

Hays, W.L. and Bennett, J.F. (1961) Multidimensional unfolding: determining configuration from complete rank order preference data. *Psychometrika*, **26**, 221-238.

Healy, M.J.R. (1986) *Matrices for Statistics*, Oxford: Clarendon Press.

Hefner, R.A. (1958) Extensions of the law of comparative judgement to discriminable and multidimensional stimuli. Doctoral dissertation, Univ. of Michigan.

Heiser, W.J. (1991) A generalized majorization method for least squares multidimensional scaling of pseudodistances that may be negative. *Psychometrika*, **56**, 7-27.

Hettmansperger, T.P. and Thomas, H. (1973) Estimation of J scales for unidimensional unfolding. *Psychometrika*, **38**, 269-284.

Hill, M.O. (1973) Reciprocal averaging: an eigenvector method of ordination. *J. Ecol.*, **61**, 237-251.

Hill, M.O. (1974) Correspondence analysis: a neglected multivariate method. *Appl. Statist.*, **23**, 340-354.

Hirschfeld, H.O. (1935) A connection between correlation and contingency. *Cambridge Phil. Soc. Proc.*, **31**, 520-524.

Horst, P. (1935) Measuring complex attitudes. *J. Social Psychol.*, **6**, 369-374.

Jackson, D.A., Somers, K.M. and Harvey, H.H. (1989) Similarity coefficients: measures of co-occurrence and association or simply measures of occurrence? *Am. Nat.*,**133**, 436-453.

Jackson, M. (1989) *Michael Jackson's Malt Whisky Companion: A Connoisseur's Guide to the Malt Whiskies of Scotland*, London: Dorling Kindersley.

Jardine, N. and Sibson, R. (1971)*Mathematical Taxonomy*. London: Wiley.

Kelly, M.J., Wooldridge, L., Hennessy, R.T., Vreuls, D., Barneby, S.F., Cotton, J.C. and Reed, J.C. (1979) Air combat maneuvering performance measurement. Williams Air Force Base, AZ: Flying Training Division, Air Force Human Resources Laboratory (NAVTRAEQUIPCEN IH 315/AFHRL-TR-79-3).

Kendall, D.G. (1971) Seriation from abundance matrices. In Hodson, F.R., Kendall, D.G. and Tătu, P. (eds.), *Mathematics in the Archaeological and Historical Sciences*, Edinburgh: Edinburgh University Press.

Kendall, D.G. (1977) On the tertiary treatment of ties. Appendix to Rivett, B.H.P., Policy selection by structural mapping. *Proc. R. Soc. Lond.*, **354**, 422-423.

Kiers, H.A.L. (1989) An alternating least squares algorithm for fitting the two- and three-way DEDICOM model and the IDIOSCAL model. *Psychometrika*, **54**, 515-521.

Kiers, H.A.L. (1991) An alternating least squares algorithm for PARA-FAC-2 and three-way DEDICOM. *Computational Statistics and Data Analysis*, **16**, 103-118.

Kiers, H.A.L. and Krijnen, W.P. (1991) An efficient algorithm for PARA-FAC of three-way data with large numbers of observation units. *Psychometrika*, **56**, 147-152.

Kiers, H.A.L., ten Berge, J.M.F., Takane, Y. and de Leeuw, J. (1990) A generalization of Takane's algorithm for DEDICOM. *Psychometrika*, **55**, 151-158.

Klahr, D. (1969) A Monte Carlo investigation of the statistical significance of Kruskal's nonmetric scaling procedure. *Psychometrika*, **34**, 319-330.

Korth, B. and Tucker, L.R. (1976) Procrustes matching by congruence coefficients. *Psychometrika*, **41**, 531-535.

Koschat, M.A. and Swayne, D.F. (1991) A weighted Procrustes criterion. *Psychometrika*, **56**, 229-239.

Kristof, W. and Wingersky, B. (1971) Generalization of the orthogonal Procrustes rotation procedure to more than two matrices. In *Proceedings, 79th Annual Convention of the American Psychological Association*, pp 81-90.

Kroonenberg, P.M. (1983) *Three-Mode Principal Components Analysis*, Leiden: DSWO Press.

Kroonenberg, P.M. and de Leeuw, J. (1980) Principal components analysis of three-mode data by means of alternating least squares algorithms. *Psychometrika*, **45**, 69-97.

Kruskal, J.B. (1964a) Multidimensional scaling by optimizing goodness-of-fit to a nonmetric hypothesis. *Psychometrika*, **29**, 1-27.

Kruskal, J.B. (1964b) Nonmetric multidimensional scaling: a numerical method. *Psychometrika*, **29**, 115-129.

Kruskal, J.B. (1971) Monotone regression: continuity and differentiability properties. *Psychometrika*, **36**, 57-62.

Kruskal, J.B. and Wish, M. (1978) *Multidimensional Scaling*. Beverly Hills, CA: Sage Publications.

Krzanowski, W.J. (1988) *Principles of Multivariate Analysis: A User's Perspective*, Oxford: Clarendon Press.

Krzanowski, W.J. (1993) Attribute selection in correspondence analysis of incidence matrices. *Appl. Statist.*, **42**, 529-541.

Langeheine, R. (1982) Statistical evaluation of measures of fit in the Lingoes-Borg Procrustean individual differences scaling. *Psychometrika*, **47**, 427-442.

Langron, S.P. and Collins, A.J. (1985) Perturbation theory for generalized Procrustes analysis. *J.R. Statist. Soc. B*, **47**, 277-284.

Lapointe, F.J. and Legendre, P. (1994) A classification of pure malt Scotch whiskies. *Appl. Statist.*, **43**, 237-257.

Lawson, W.J. and Ogg, P.J. (1989) Analysis of phenetic relationships among populations of the avian genus Batis (Platysteirinae) by means of cluster analysis and multidimensional scaling. *Biom. J.*, **31**, 243-254.

Lee, S.Y. (1984) Multidimensional scaling models with inequality and equality constraints. *Commun. Statist.-Simula. Computa.*, **13**, 127-140.

Lee, S.Y. and Bentler, P.M. (1980) Functional relations in multidimensional scaling. *British Journal of Mathematical and Statistical Psychology*, **33**, 142-150.

Levine, D.M. (1978) A Monte Carlo study of Kruskal's variance based measure on stress. *Psychometrika*, **43**, 307-315.

Lingoes, J.C. and Borg, I. (1976) Procrustean individual differences scaling: PINDIS. *Journal of Marketing Research*, **13**, 406-407.

Lingoes, J.C. and Borg, I. (1977) Procrustean individual differences scaling: PINDIS. *Sozial Psychologie*, **8**, 210-217.

Lingoes, J.C. and Borg, I. (1978) A direct approach to individual differences scaling using increasingly complex transformations. *Psychometrika*, **43**, 491-519.

Lingoes, J.C. and Roskam, E.E. (1973) A mathematical and empirical study of two multidimensional scaling algorithms. *Psychometrika Monograph Supplement*, **38**.

Lissitz, R.W., Schönemann, P.H. and Lingoes, J.C. (1976) A solution to the weighted Procrustes problem in which the transformation is in agreement with the loss function. *Psychometrika*, **41**, 547-550.

MacCallum, R.C. (1976) Effects on INDSCAL of non-orthogonal perceptions of object space dimensions. *Psychometrika*, **41**, 177-188.

MacCallum, R.C. (1977a) Effects of conditionality on INDSCAL and ALSCAL weights. *Psychometrika*, **42**, 297-305.

MacCallum, R.C. (1977b) A Monte Carlo investigation of recovery of structure by ALSCAL. *Psychometrika*, **42**, 401-428.

MacCallum, R.C. (1978) Recovery of structure in incomplete data by ALSCAL. *Psychometrika*, **44**, 69-74.

MacCallum, R.C. and Cornelius III, E.T. (1977) A Monte Carlo investigation of recovery of structure by ALSCAL. *Psychometrika*, **42**, 401-428.

Mardia, K.V. (1978) Some properties of classical multidimensional scaling. *Commun. Statist. Theor. Meth.*, **A7**, 1233-1241.

Mardia, K.V., Kent, J.T. and Bibby, J.M. (1979) *Multivariate Analysis*, London: Academic Press.

McElwain, D.W. and Keats, J.A. (1961) Multidimensional unfolding: some geometrical solutions. *Psychometrika*, **26**, 325-332.

Mead, A. (1992) Review of the development of multidimensional scaling methods. *The Statistician*, **41**, 27-39.

Messick, S.M. and Abelson, R.P. (1956) The additive constant problem in multidimensional scaling. *Psychometrika*, **21**, 1-15.

Mooijaart, A. and Commandeur, J.J.F. (1990) A general solution of the weighted orthonormal Procrustes problem. *Psychometrika*, **55**, 657-663.

New Geographical Digest (1986), London: George Philip.

Nishisato, S. (1980) *Analysis of Categorical Data: Dual Scaling and its Applications*, Totonto: University of Toronto Press.

Pack, P. and Jolliffe, I.T. (1992) Influence in correspondence analysis. *Appl. Statist.*, **41**, 365-380.

Pan, G. and Harris, D.P. (1991) A new multidimensional scaling technique based upon associations of triple objects – Pijk and its application to the analysis of geochemical data. *Mathematical Geology*, **6**, 861-886.

Peay, E.R. (1988) Multidimensional rotation and scaling of configurations to optimal agreement. *Psychometrika*, **53**, 199-208.

Plackett, R.L. (1981) *The Analysis of Categorical Data*, London: Griffin.

Polzella, D.J. and Reid, G.R. (1989) Multidimensional scaling analysis of simulated air combat maneuvering performance data. *Aviat. Space, Environ. Med.*, **60**, 141-144.

Poste, L.M. and Patterson, C.F. (1988) Multidimensional scaling – sensory analysis of yoghurt. *Can. Inst. Food Sci. Technol. J.*, **21**, 271-278.

Ramsay, J.O. (1977) Maximum likelihood estimation in multidimensional scaling. *Psychometrika*, **42**, 241-266.

Ramsay, J.O. (1978a) Confidence regions for multidimensional scaling analysis. *Psychometrika*, **43**, 145-160.

Ramsay, J.O. (1978b) *MULTISCALE: Four Programs of Multidimensional Scaling by the Method of Maximum Likelihood*. Chicago: International Educational Services.

Ramsay, J.O. (1980) Some small sample results for maximum likelihood estimation in multidimensional scaling. *Psychometrika*, **45**, 141-146

Ramsay, J.O. (1982) Some statistical approaches to multidimensional scaling data. *J. R. Statist. Soc.*, **A 145**, 285-312.

Richardson, M. and Kuder, G.F. (1933) Making a rating scale that measures. *Personnel J.*, **12**, 36-40.

Ripley, B.D. (1981) *Spatial Statistics*. New York: Wiley.

Rivett, B.H.P. (1977) Policy selection by structural mapping. *Proc. R. Soc. Lond.*, **354**, 407-423.

Roberts, G., Martyn, A.L., Dobson, A.J. and McCarthy, W.H. (1981) Tumour thickness and histological type in malignant melanoma in New South Wales, Australia. 1970-76. *Pathology*, **13**, 763-770.

Ross, J. and Cliff, N. (1964) A generalization of the interpoint distance model. *Psychometrika*, **29**, 167-176.

Saito, T. (1978) The problem of the additive constant and eigenvalues in metric multidimensional scaling. *Psychometrika*, **43**, 193-201.

Sammon, J.W. (1969) A nonlinear mapping for data structure analysis. *IEEE Transactions on Computers*, **18**, 401-409.

Schiffman, S.S, Reynolds, M.L. and Young, F.W. (1981) *Introduction to Multidimensional Scaling: Theory, Methods and Applications*, New York: Academic Press.

Schobert, R. (1979) *Die Dynamisierung komplexer Marktmodelle mit Hilfe von Verfahren der mehrdimensionalen Skalierung*, Berlin: Duncker and Humblot.

Schoenberg, I.J. (1935) Remarks to Maurice Fréchet's article "Sur la définition axiomatique d'une classe d'espaces vectoriels distanciés applicables vectoriellement sur l'espace de Hilbert". *Ann. Math.*, **36**, 724–732.

Schönemann, P.H. (1966) A generalized solution of the orthogonal Procrustes problem. *Psychometrika*, **31**, 1-10.

Schönemann, P.H. (1970) On metric multidimensional unfolding. *Psychometrika*, 35, 349-366.

Schönemann, P.H. and Carroll, R.M. (1970) Fitting one matrix to another under choice of a central dilation and a rigid motion. *Psychometrika*, **35**, 245-256.

Shepard, R.N. (1962a) The analysis of proximities: multidimensional scaling with an unknown distance function I. *Psychometrika*, **27**, 125-140.

Shepard, R.N. (1962b) The analysis of proximities: multidimensional scaling with an unknown distance function II. *Psychometrika*, **27**, 219-246.

Sherman, C.R. (1972) Nonmetric multidimensional scaling: a Monte Carlo study of the basic parameters. *Psychometrika*, **37**, 323-355.

Sibson, R. (1978) Studies in the robustness of multidimensional scaling: Procrustes statistics. *J. R. Statist. Soc.*, **B 40**, 234-238.

Sibson, R. (1979) Studies in the robustness of multidimensional scaling; perturbational analysis of classical sacaling. *J. R. Statist. Soc.*, **B 41**, 217-229.

Sibson, R., Bowyer, A. and Osmond, C. (1981) Studies in the robustness of multidimensional scaling: Euclidean models and simulation studies. *J. Statist. Comput. Simul.*, **13**, 273-296.

Smith, N.J. and Iles, K. (1988) A graphical depiction of multivariate similarity among sample plots. *Can. J. For. Res.*, **18**, 467-472.

Sneath, P.H.A. and Sokal, R.R. (1973) *Numerical Taxonomy*. San Francisco: W.H. Freeman and Co.

Snijders, T.A.B., Dormaar, M., van Schuur, W.H., Dijkman-Caes, C. and Driessen, G. (1990) Distribution of some similarity coefficients for dyadic binary data in the case of associated attributes. *Journal of Classification*, **7**, 5-31.

Spaeth, H.J. and Guthery, S.B. (1969) The use and utility of the monotone criterion in multidimensional scaling. *Multivariate Behavioral Research*, **4**, 501-515.

Spence, I. (1970) Local minimum solutions in nonmetric multidimensional scaling. *Proc. of the Soc. Stats. Section of the American Statist. Assoc.*, **13**, 365-367.

Spence, I. (1972) A Monte Carlo evaluation of three nonmetric multidimensional scaling algorithms. *Psychometrika*, **37**, 461-486.

Spence, I. and Domoney, D.W. (1974) Single subject incomplete designs for nonmetric multidimensional scaling. *Psychometrika*, **39**, 469-490.

Spence, I. and Lewandowsky, S. (1989) Robust multidimensional scaling. *Psychometrika*, **54**, 501-513.

Spence, I. and Ogilvie, J.C. (1973) A table of expected stress values for random rankings in nonmetric multidimensional scaling. *Multivariate Behavioral Research*, **8**, 511-517.

Stenson, H.H. and Knoll, R.L. (1969) Goodness of fit for random rankings in Kruskal's nonmetric scaling procedure. *Psychological Bulletin*, **71**, 122-126.

Takane, Y. (1978a) A maximum likelihood method for nonmetric multidimensional scaling: I. The case in which all empirical pairwise orderings are independent – theory. *Japanese Psychological Research*, **20**, 7-17.

Takane, Y. (1978b) A maximum likelihood method for nonmetric multidimensional scaling: I. The case in which all empirical pairwise orderings are independent – evaluation. *Japanese Psychological Research*, **20**, 105-114.

Takane, Y. (1981) Multidimensional successive categories scaling: a maximum likelihood method. *Psychometrika*, **46**, 9-28.

Takane, Y., Young, F.W. and de Leeuw, J.(1977) Nonmetric individual differences multidimensional scaling: an alternating least squares method with optimal scaling features. *Psychometrika*, **42**, 7-67.

ten Berge, J.M.F. (1977) Orthogonal Procrustes rotation for two or more matrices. *Psychometrika*, **42**, 267-276.

ten Berge, J.M.F. (1983) A generalization of Verhelst's solution for a constrained regression problem in ALSCAL and related MDS-algorithms. *Psychometrika*, **48**, 631-638.

ten Berge, J.M.F. and Knol, D.L. (1984) Orthogonal rotations to maximal asgreement for two or more matrices of different column orders. *Psychometrika*, **49**, 49-55.

ten Berge, J.M.F. and Nevels, K. (1977) A general solution to Mosier's oblique Procrustes problem. *Psychometrika*, **42**, 593-600.

ten Berge, J.M.F., de Leeuw, J. and Kroonenberg, P.M. (1987) Some additional results on principal components analysis of three-mode data by means of alternating least squares algorithms. *Psychometrika*, **52**, 183-191.

Tenenhaus, M. and Young, F.W. (1985) An analysis and synthesis of multiple correspondence analysis, optimal scaling, dual scaling, homogeneity analysis and other methods for quantifying categorical multivariate data. *Psychometrika*, **50**, 91-119.

Ter Braak, C.J.F. (1992) Multidimensional scaling and regression. *Statistica Applicata*, **4**, 577-586.

Tijssen, R.J.W. and Van Raan, A.F.J. (1989) Mapping co-word structures: a comparison of multidimensional scaling and leximappe. *Scientometrics*, **15**, 283-295.

Tong, S.T.Y. (1989) On nonmetric multidimensional scaling ordination and interpretation of the matorral vegetation in lowland Murcia. *Vegetatio*, **79**, 65-74.

Torgerson, W.S. (1952) Multidimensional scaling: 1. Theory and method. *Psychometrika*, **17**, 401-419.

Torgerson, W.S. (1958) *Theory and Method of Scaling*, New York: Wiley.

Tucker, L.R. (1951) A method for synthesis of factor analytic studies. Personnel Research Section Report No. 984, Department of the Army, Washington, DC.

Tucker, L.R. (1966) Some mathematical notes on three-mode factor analysis. *Psychometrika*, **31**, 279-311.

Tucker, L.R. (1972) Relations between multidimensional scaling and three-mode factor analysis. *Psychometrika*, **37**, 3-27.

Tucker, L.R. and Messick, S. (1963) An individual differences model for multidimensional scaling. *Psychometrika*, **28**, 333-367.

Van der Heijden, P.G.M. and de Leeuw, J. (1985) Correspondence analysis used complementary to loglinear analysis. *Psychometrika*, **50**, 429-447.

Van der Heijden, P.G.M. and Meijerink, F. (1989) Generalized correspondence analysis of multi-way contingency tables and multi-way (super-) indicator matrices. In Coppi, R. and Bolasco, S. (eds.), *Multiway Data Analysis*. Amsterdam: North-Holland, pp 185-202.

Van der Heijden, P.G.M and Worsley, K.J. (1988) Comment on "Correspondence analysis used complementary to loglinear analysis". *Psychometrika*, **53**, 287-291.

Van der Heijden, P.G.M., de Falguerolles, A. and de Leeuw, J. (1985) A combined approach to contingency table analysis using correspondence analysis and log-linear analysis. *Appl. Statist.*, **38**, 249-292.

Verhelst, N.D. (1981) A note on ALSCAL: the estimation of the additive constant. *Psychometrika*, **46**, 465-468.

Wagenaar, W.A. and Padmos, P. (1971) Quantitative interpretation of stress in Kruskal's multidimensional scaling technique. *Br. J. Math. Statist. Psychol.*, **24**, 101-110.

Weeks, D.G. and Bentler, P.M. (1982) Restricted multidimensional scaling models for asymmetric proximities. *Psychometrika*, **47**, 201-208.

Winsberg, S. and Carroll, J.D. (1989a) A quasi-nonmetric method for multidimensional scaling of multiway data via a restricted case of an extended INDSCAL model. In Coppi, R. and Bolasco, S. (eds.), *Multiway Data Analysis*, Amsterdam: North Holland.

Winsberg, S. and Carroll, J.D. (1989b) A quasi-nonmetric method for multidimensional scaling of multiway data via a restricted case of an extended Euclidean model. *Psychometrika*, **54**, 217-229.

Winsberg, S. and De Soete, G. (1993) A latent class approach to fitting the weighted Euclidean model, CLASCAL. *Psychometrika*, **58**, 315-330.

Wish, M. and Carroll, J.D. (1982) Theory of multidimensional scaling. In

Krishnaiah, P.R. and Kanal, L.N. (eds.), *Handbook of Statistics, vol 2*, Amsterdam: North Holland, pp 317-345.

Wish, M., Deutsch, M. and Biener, L. (1972) Differences in perceived similarity of nations. In Romney, A.K., Shepard, R.N. and Nerlove, S.B. (eds.), *Theory and Applications in the Behavioural Sciences, Vol. 2*, New York: Seminar Press, pp 289-313.

Young, F.W. (1987) *Multidimensional Scaling: History, Theory and Applications*, Hamer, R.M. (ed.), Hillsdale, NJ: Lawrence Erlbaum.

Young, F.W. and Cliff, N.F (1972) Interactive scaling with individual subjects. *Psychometrika*, **37**, 385-415.

Young, F.W. and Null, C.H. (1978) Multidimensional scaling of nominal data: the recovery of metric information with ALSCAL. *Psychometrika*, **43**, 367-379.

Young, F.W., de Leeuw, J. and Takane, Y. (1976) Regression with qualitative and quantitative variables: an alternating least squares method with optimal scaling features. *Psychometrika*, **41**, 505-529.

Young, F.W., Takane, Y. and Lewyckyj, R. (1978) Three notes on ALSCAL. *Psychometrika*, **43**, 433-435.

Young, G. and Householder, A.S. (1938) Discussion of a set of points in terms of their mutual distances. *Psychometrika*, **3**, 19–22.

Zegers, F.E. (1986) A family of chance-corrected association coefficients for metric scales. *Psychometrika*, **51**, 559-562.

Zegers, F.E. and ten Berge, J.M.F. (1985) A family of association coefficients for metric scales. *Psychometrika*, **50**, 17-24.

Zinnes, J.L. and Griggs, R.A. (1974) Probabilistic, multidimensional unfolding analysis. *Psychometrika*, **39**, 327-350.

Zinnes, J.L. and MacKay, D.B. (1983) Probabilistic multidimensional scaling: complete and incomplete data. *Psychometrika*, **48**, 27-48.

Author index

Subject index

H61. K75 Wilson